Energy-Related Material Flow Simulation in Production and Logistics

Sigrid Wenzel · Markus Rabe ·
Steffen Strassburger · Christoph von Viebahn
Editors

Energy-Related Material Flow Simulation in Production and Logistics

Springer

Editors
Sigrid Wenzel
Fachgebiet Produktionsorganisation und Fabrikplanung, Fachbereich Maschinenbau
Universität Kassel
Kassel, Germany

Steffen Strassburger
Fachgebiet Informationstechnik in Produktion und Logistik
Fakultät für Maschinenbau
Technische Universität Ilmenau
Ilmenau, Germany

Markus Rabe
IT in Produktion und Logistik
TU Dortmund University
Dortmund, Germany

Christoph von Viebahn
Fakultät Wirtschaft und Informatik
Institute for Applied Data Science Hanover
Hochschule Hannover
Hannover, Niedersachsen, Germany

ISBN 978-3-031-34217-2 ISBN 978-3-031-34218-9 (eBook)
https://doi.org/10.1007/978-3-031-34218-9

This Springer imprint is published by the registered company Springer Nature Switzerland AG
The registered company address is: Gewerbestrasse 11, 6330 Cham, Switzerland

Paper in this product is recyclable.

Preface

Since the beginning of the industrialization age, engineers have aimed to increase productivity and reduce costs. In the last decades, customer orientation has steadily gained importance and, thus, short and reliable delivery times have become a competing target, combined with a trend to mass customization. Currently, however, sustainability aspects are moving into the focus of customers and enterprises. With respect to production and logistics tasks, this mainly affects the consumption of energy and, in consequence, the emission of greenhouse gas (GHG). This trend has been amplified by the dramatic increase in energy costs after the outbreak of the war in Ukraine, and mirrored by national and international taxes and regulations, such as the European Sustainability Reporting Standards (ESRS). The scheduled Corporate Sustainability Reporting Directive (CSRD) will have distinct implications on the annual audit and liquidity of companies.

Methodologies to face these challenges range from the specific acquisition of data on energy consumption via the allocation of particular production processes and the experimental planning of improvement up to the simulation of how to integrate an increasing percentage of renewable energy into current production and logistics processes. The use cases of the book promote to apply methodologies that help to comply with these upcoming challenges.

The first applications of material flow simulation have already been reported for about 50 years. In the last 40 years, simulation has been successfully introduced to analyze and improve first the material flow and later the related information flow, enabling engineers to gain deep insights into the behavior of complex modern production and logistics systems. Sometimes, energy-related aspects have been considered, but in most cases indirectly, e.g., reducing the runtime of equipment and only by this measure decreasing the energy consumption. However, the importance of respecting energy in the processes has become more and more urgent, and the pressure to reduce the environmental footprint of production and logistics systems will intensify in the upcoming decade. Therefore, enterprises have started to integrate their consumption of energy into their planning processes much more frequently than before, even constructing feedback loops, e.g., from energy control to production control. This receives additional attention for the increasing use of renewable, but less reliable, energy sources. Care must be taken to establish processes that aim to use energy when it is available. As an example, many industrial processes like melting or coating have significant energy demands, but could vary the point of time of its consumption within specific limits, leading to very high complexity.

Simulation is the technology of choice for the analysis of such complex interconnected systems. Nevertheless, there is no satisfying overview of the current approaches and applications of considering energy for production and logistics simulation. The section "Simulation in Production and Logistics (SPL)" of the Association for Simulation in the German-speaking Area (Germany, Switzerland, and Austria) (ASIM) responded to the importance of these developments by founding the "Workgroup on the Investigation of Energy-related Influences in SPL" in 2014. It has gathered an extensive and structured collection of relevant works to shed light on the findings of various groups or organizations as well as on knowledge gaps. Major results are now published in this book, which, therefore, is also registered as ASIM Proceedings No. 182. In Part I, the book introduces the approaches to model energy-related aspects in the simulation of production and logistics systems that are available today, discusses the construction and application of energy-specific performance indicators, and analyzes the input information that needs to be acquired before implementing suitable models. On this basis, the technical solutions are introduced.

Regarding practical implementation and illustration, Part II of the book is divided into six chapters, each dedicated to one application field, such as automotive, electronics, and transportation. In each of these chapters, written by related experts, the specific performance indicators and required data are introduced, challenges to the conceptual modeling explained with their solution approaches, and, finally, several examples given for the application of these approaches. Thus, these chapters can support the engineers of the related domains for understanding the scope and tasks for a suitable simulation model, and to achieve an estimate of the effort that it might require and the benefits it could raise.

The editors express their gratitude to all members of the ASIM working group for the investigation of energy-related influences in SPL for the many discussions, the evaluation of numerous articles and papers, and the many years of commitment to this topic. Special thanks are addressed to the many authors of this book, who invested huge effort and care to provide the readers with an exciting and informative experience.

Kassel, Germany — Sigrid Wenzel
Dortmund, Germany — Markus Rabe
Ilmenau, Germany — Steffen Strassburger
Hannover, Germany — Christoph von Viebahn
December 2022

Summary

Content Description for the Publisher

Material flow simulation has been successfully applied for about 50 years to analyze and improve first the material flow and later the related information flow, enabling engineers to gain deep insights into the behavior of complex modern production and logistics systems. Sometimes, energy-related aspects have been considered, but in most cases indirectly, e.g., reducing the runtime of equipment and in consequence decreasing the energy consumption.

However, in the last decade, the importance of respecting energy in the processes has become more and more important, and the pressure to reduce the environmental footprint of production and logistics systems will intensify in the upcoming decade. Therefore, enterprises have started to integrate the use of energy into their planning processes much more frequently than before, even constructing feedback loops, e.g., from energy control to production control. This receives additional attention with the increasing use of renewable, but less reliable, energy sources. Care must be taken to establish processes that aim to use energy when it is available. As an example, many industrial processes like melting or coating have significant energy demands, but could vary the point of time of its consumption within specific limits, leading to very high complexity.

Simulation is the technology of choice for such complex interconnected systems. Nevertheless, there is no satisfying overview of the current approaches and applications of considering energy for production and logistics simulation. To fill this gap, this book introduces in Part I the approaches to model energy-related aspects in this context that are available today, discusses the construction and application of energy-specific performance indicators, and analyzes the input information that needs to be acquired before implementing suitable models. On this basis, the technical solutions are introduced.

For the practical implementation and illustration, Part II of the book is divided into six chapters, each related to one application field, such as automotive, perishables, and transportation. In each of these chapters, written by related experts, the specific performance indicators and required data are introduced, challenges to the conceptual modeling explained with their solution approaches, and, finally, examples given for the application of these approaches. Thus, these chapters can support the engineers of the related domains to understand the scope and tasks for a suitable simulation model, and to achieve an estimate of the effort it might require and the benefits it could raise.

Target Groups

The book is targeted to engineers and scientists investigating energy use aspects that are connected to the material flow in production and logistics systems in a broad sense, including any kinds of transport, buffering, and the control of interrelated processes. The provided state of the art helps engineers to select and understand modeling techniques that are suitable for their specific tasks. It also forms a sound base for further scientific research, and can be used in advanced teaching, e.g., for university masters, to educate engineers in this field with massively growing importance: Few engineers studying today will not be concerned with energy efficiency topics in their business career. For the practitioners, the chapters in Part II of the book give even more specific hints on how to handle typical energy-related questions in the specific branches, and also provide an illustration of possibilities that engineers can take as samples or as a stimulus for their own work. Finally, managers who are responsible for decisions in the improvement of energy use and the application of simulation find precious samples and can improve their understanding of the technology's benefits and challenges.

Contents

Editors and Contributors

About the Editors

Sigrid Wenzel is a Professor and Head of the Department of Production Organization and Factory Planning, University of Kassel. In addition to this, she is a Board Director of the Arbeitsgemeinschaft Simulation (ASIM), Spokesperson of the ASIM Section Simulation in Production and Logistics (SPL) as well as of the workgroup on the Investigation of Energy-related Influences in SPL, member of the Advisory Board of the Association of German Engineers Society of Production and Logistics (VDI-GPL), and Head of the Committee Modeling and Simulation of the VDI-GPL.

Markus Rabe is Full Professor of IT in Production and Logistics (ITPL) at the Technical University Dortmund. Until 2010, he had been with Fraunhofer IPK in Berlin as head of the Corporate Logistics and Processes department, head of the Central IT department, and a member of the Institute Direction Circle. His research focus is on information systems for supply chains, production planning, and simulation. Markus Rabe is Vice Chair of the "Simulation in Production and Logistics" group of the simulation society ASIM, Member of the Editorial Board of the *Journal of Simulation*, member of several conference program committees, has chaired the ASIM SPL conference in 1998, 2000, 2004, 2008, and 2015, is Local Chair of the WSC'2012 in Berlin, and Proceedings Chair of the WSC'18 and WSC'19. More than 220 publications and editions report from his work.

Steffen Strassburger is a Professor at the Ilmenau University of Technology and Head of the Group for Information Technology in Production and Logistics. Previously, he was head of the "Virtual Development" department at the Fraunhofer Institute in Magdeburg, Germany, and a researcher at the DaimlerChrysler Research Center in Ulm, Germany. He holds a Doctoral and a Diploma degree in Computer Science from the University of Magdeburg, Germany. His research interests include distributed simulation, automatic simulation model generation, and general interoperability topics within the digital factory and Industry 4.0 context.

Christoph von Viebahn is a Professor for Business Informatics at the University of Applied Sciences, Hannover. He is specialized in Supply Chain Management, Business Administration, and Dynamic Computer Simulation. He is Regional Group Chairperson for Lower Saxony by the Bundesvereinigung Logistik (BVL) since 2016 and Member of the Urban Logistics working group since 2017. His research portfolio has been decorated with the research award of Lower Saxony in 2022.

Contributors

Marvin Auf der Landwehr Fakultät Wirtschaft und Informatik, Institute for Applied Data Science Hanover, Hochschule Hannover, Hannover, Germany

Lukas Bank Berlin, Germany

Karsten Beyer SimPlan AG—Dresden Branch, Dresden, Germany

Jorge Luis Chicaiza-Vaca Pichincha Chamber of Freight Transportation, Quito, Ecuador

Antal Dér Sustainable Manufacturing and Life Cycle Engineering, Institute of Machine Tools and Production Technology (IWF), Technische Universität Braunschweig, Braunschweig, Germany

Johannes Dettelbacher Reckendorf, Germany

Javier Faulin Institute of Smart Cities, Department of Statistics, Computer Science, and Mathematics, Public University of Navarra, Pamplona, Spain

Christian Fikar Faculty VII for Life Sciences: Food, Nutrition and Health, Chair of Food Supply Chain Management, University of Bayreuth, Kulmbach, Germany

Kai Gutenschwager Institute of Information Engineering, Ostfalia University of Applied Sciences, Wolfenbüttel, Germany

Björn Johansson Department of Industrial and Materials Science, Chalmers University of Technology, Gothenburg, Sweden

Michael E. Kuhl Industrial and Systems Engineering Department, Rochester Institute of Technology, Rochester, NY, USA

Marc Münnich Fraunhofer Institute for Machine Tools and Forming Technology IWU, Chemnitz, Germany

Tim Peter Großweil, Germany

Markus Rabe Fachgebiet IT in Produktion und Logistik, TU Dortmund University, Dortmund, Germany

Martin Rösch Bayernwerk Netz GmbH, Regensburg, Germany

Wolfgang Schlüter Nürnberg, Germany

Cedric Schultz TARGUS Management Consulting AG, Ratingen, Germany

Adrian Serrano-Hernandez Institute of Smart Cities, Department of Statistics, Computer Science, and Mathematics, Public University of Navarra, Pamplona, Spain

Thomas Sobottka Fraunhofer Austria Research GmbH, Vienna, Austria

Kristina Sokoll Ingolstadt, Germany

Johannes Stoldt ABB Stotz-Kontakt GmbH, Heidelberg, Germany

Steffen Strassburger Technische Universität Ilmenau, Fakultät für Maschinenbau, Fachgebiet Informationstechnik in Produktion und Logistik, Ilmenau, Germany

Sebastian Thiede Chair of Manufacturing Systems, Department of Design, Production and Management, Faculty of Engineering Technology, University of Twente, Enschede, The Netherlands

Christoph von Viebahn Fakultät Wirtschaft und Informatik, Institute for Applied Data Science Hanover, Hochschule Hannover, Hannover, Germany

Part I
General Considerations

Chapter 1
Classification, Input Data, and Key Performance Indicators

Markus Rabe, Johannes Stoldt, Steffen Strassburger, and Christoph von Viebahn

Abstract Simulation is a well-known technology for production and logistics, especially for the planning of new systems and the examination of ideas to optimize existing ones. In the past, the main target of such studies has been costs of equipment and personnel, but the continuously stricter view on consumption of energy has shifted this focus towards the analysis of energy consumption and emission of greenhouse gas. In some cases this might be straightforward, e.g., when the resulting production hours can just be multiplied with energy consumption per hour. Many cases, however, are far more complicated and can only be sufficiently analyzed when the detailed dynamics of energy consumption are already considered in the simulation model. Thus, a number of different approaches exist to model energy aspects in simulation models, depending on the goal of the investigation and the kind of production or logistics process. This chapter classifies these approaches in a morphological box and explains the details of the related categories. Furthermore, it discusses the requirements to input data that arise when simulation models are amended with energy components, and discusses the additional results that can be gained from such models.

M. Rabe (✉)
Technische Universität Dortmund, Fachgebiet IT in Produktion und Logistik, Leonhard-Euler-Straße 5, 44227 Dortmund, Germany
e-mail: markus.rabe@tu-dortmund.de

J. Stoldt
ABB Stotz-Kontakt GmbH, Eppelheimer Str. 82, 69123 Heidelberg, Germany
e-mail: johannes.stoldt@de.abb.com

S. Strassburger
Technische Universität Ilmenau, Fakultät für Maschinenbau, Fachgebiet Informationstechnik in Produktion und Logistik, Max-Planck-Ring 12, 98693 Ilmenau, Germany
e-mail: steffen.strassburger@tu-ilmenau.de

C. von Viebahn
Fakultät Wirtschaft und Informatik, Institute for Applied Data Science Hanover, Hochschule Hannover, Ricklinger Stadtweg 120, 30459 Hannover, Germany
e-mail: christoph-von.viebahn@hs-hannover.de

S. Wenzel et al. (eds.), *Energy-Related Material Flow Simulation in Production and Logistics*, https://doi.org/10.1007/978-3-031-34218-9_1

This chapter introduces the theoretical background for the integration of energy aspects within simulation models for production and logistics. While discrete event simulation (DES) is well studied and commonly applied in production and logistics, the usage of DES to evaluate energy-related aspects of the studied systems is a rather new topic that has gained increasing importance. The range of energy-related aspects that may be of interest is just wide. Straightforward items include the electric energy and power consumption of the systems under investigation, or, more generally speaking, the demand side of the production or logistic system. Nevertheless, every production and logistic system also has further energy-related impact that one might wish to study, e.g., waste heat or emissions such as carbon dioxide.

To provide a better understanding of these aspects, this chapter discusses the following content: Sect. 1.1 gives a short insight on the method of literature collection and summarizes the major past activities of the ASIM working group in this field. Section 1.2 presents a classification of the different approaches for the integration of energy aspects and illustrates the multiplicity of available options. Section 1.3 discusses additional input information required for the depiction of energy aspects. Section 1.4 introduces key performance indicators that can be extracted and analyzed when energy aspects have been integrated into simulation models.

1.1 Identification of Related Publications

In this first section, the background of the underlying literature and the research methods to acquire the current state of the art are explained. The section "Simulation in Production and Logistics (SPL)" of the Association for Simulation in the German-speaking Area (Germany, Switzerland, and Austria) (ASIM) founded a specific "Workgroup on the Investigation of Energy-related Influences in SPL" in 2014. The members of this workgroup have conducted an extensive and structured collection of relevant publications (Wenzel et al. 2017; Uhlig et al. 2018; Poeting et al. 2019; Stoldt et al. 2021). Dependencies among key performance indicators and the level of detail for modeling energy aspects have been analyzed and their effects on specific requirements for simulation input data have been studied by the workgroup. Furthermore, particularities that depend on the major scope, as logistics or manufacturing, have been investigated (Poeting et al. 2019), with manufacturing forming the majority of the found cases.

Various methodological approaches for literature reviews can be distinguished, which primarily differ in the way they try to incorporate quantitative data about the identified publications. Meta-analyses exploit several previous studies and mathematically integrate their results. Systematic reviews focus on the effects of specific indicators when summarizing existing quantitative studies. Scoping reviews cover all kinds of relevant publications, e.g., to identify the coverage of existing knowledge as well as recognizable gaps (Sturma et al. 2016). Having in mind the workgroup's intention to understand available competencies and white fields in the simulation including energy aspects, the methodology for scoping reviews has been selected as the most suitable method, as proposed by Colquhoun et al. (2014).

The general process following scoping reviews has been described by Uhlig et al. (2018) and is summarized in Fig. 1.1. The depicted list of keywords is not applied automatically but manually, in order to enable approximate matches filtering for relevant work. As of May 2022, more than 200 publications from journals (e.g., Journal of Simulation, Simulation, Simulation Practice and Theory, CIRP Annals Manufacturing Technology, Journal of Cleaner Production) and conference proceedings (e.g., Winter Simulation Conference, ASIM Conference Production and Logistics, Procedia CIRP) as well as other publication channels (e.g., thesis works) have been identified and added to the scoping review's body of literature. The analyses finally led to a set of characteristics for energy-related simulation models, which are summarized in a morphological box and explained in the following sections.

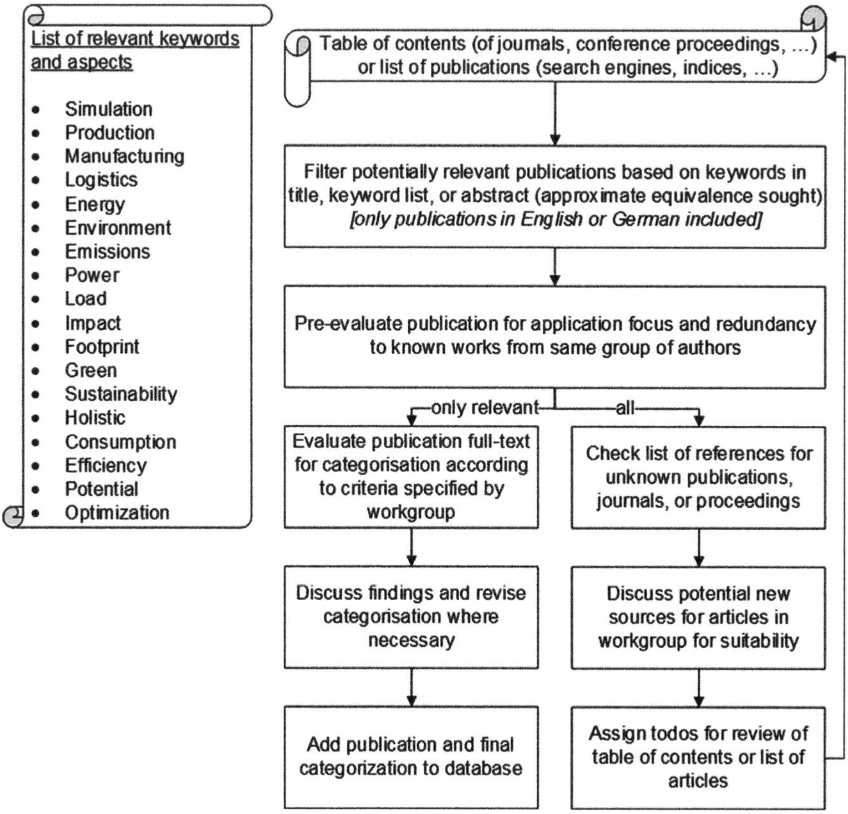

Fig. 1.1 Methodology of the scoping review (Stoldt et al. 2021)

1.2 Approaches for Integration of Energy Aspects

When trying to investigate energy-related aspects in conjunction with the DES models commonly used in production and logistics, different options exist. An obvious—but not necessarily the best—choice might be to augment an existing general simulation model with extensions depicting energy-related aspects. The decision for a specific approach depends on the available input data, the desired output, and the required accuracy. Therefore, in this section, the available options are systematically analyzed, including their individual advantages and disadvantages, and prescriptive guidance provided on the selection of suitable approaches based on a given context and intended application. To describe the different options systematically, a morphological box is proposed with several dimensions illustrating the possible variants. This morphological box is also the basis for illustrating the different applications discussed in Part II of this book.

1.2.1 Morphological Box

The dimensions of the morphological box assess the methodological aspects of modeling and simulation (M&S), objectives and result data, additional input data, as well as the industry sectors and manufacturing structures that the case studies from Part II of this book relate to. Each dimension of the morphological box is subdivided into different criteria that shape this dimension. Each criterion can have different values and sometimes even sub-values. Overall, the morphological box can be used to systematically analyze and classify a certain M&S solution to a domain problem. The other way round, it can help in finding and defining a solution to a certain domain problem. Starting with the problem domain, the desired result data and objectives, and the possibly available input data, one can select an appropriate methodological approach for modeling by looking at comparable solutions already classified within the morphological box. The different dimensions of the morphological box are introduced in Sects. 1.2.2–1.2.5. The complete box is shown in Fig. 1.2.

1.2.2 Dimension 1: Methodological Aspects of M&S

When integrating energy aspects into simulation models for production and logistics, a suitable methodological modeling approach is required. Therefore, the dimension "Methodological Aspects of M&S" classifies the different options that have been proposed in the scientific literature.

Modeling approach for energy aspects The first criterion of this dimension is the actual modeling approach chosen for the energy aspects of the models. While simulation models for production and logistics are traditionally based on discrete

Methodological Aspects of M&S	Modeling approach for energy aspects	Discrete		Continuous		Machine Learning/Artificial Intelligence		
		Means or status-based	Time series	Differential equations	System Dynamics	Deep Neural Networks	Decision trees	Other
	Timing of the energy evaluation	Concurrent			Retrospective			
	Architectural approach	Energy evaluation integrated in the simulation model	Simulation tool coupled with evaluation tool	Separate simulation models		Retrospective evaluation in suitable tool (e.g., spreadsheet)		
				Online coupling	Offline coupling			
Objectives and Result Data	Objective for simulation of energetic aspects	Dimensioning and Design		Optimization of energy consumption and costs	Load peak avoidance / reduction	Forecast of the energy demand		Other
		of the energy infrastructure	of the production / logistics system					
	Relevant result data	Energy consumption	Energy costs	Power requirement	Emissions & environmental impacts	Energy-related key performance indicators		Other
				Peak load / Average load		Output quantity	Level of supply	
Additional Input Data	For modeling energetic aspects	Nominal performance data (time-dependent/time-independent)				Measurement data of the real system	Physical behavior models	Other
		Machines & Robots	Transport system	Ware-house	Other Facilities			
	For the evaluation	Energy sources		Energy prices (and models)		Emission equivalents	Other	

Fig. 1.2 Morphological box of energy integration aspects

event simulation, modeling approaches for energy aspects can be either discrete or continuous. Such approaches can even take an entirely different modeling form, e.g., by applying models based on machine learning (ML) or artificial intelligence (AI).

Discrete approaches are especially suitable when a status-based approach for depicting energy aspects is sufficient. As an example, this might be the approach of choice for situations where the mean energy consumption can be matched to certain machine states (Stoldt et al. 2016). Then, the energy evaluation would be able to accumulate energy consumption for all machines and resources in the simulation models. The granularity of this approach is limited, as there is no differentiation of energy consumption within a certain machine state. Still, for many applications such an approach could be sufficient.

Industrial Sectors and Manufacturing Type

Industrial Sectors	Manufacturing				Transportation	Retail	Perishables	Renewables
	Automotive		Electronics	General				
	Vehicle construction	Components / Supply parts						

Manufacturing type	Make-to-order production	Series production	Variety production	Mass production	None (trade, logistics...)
Manufacturing structure	Flow production	Workshop production	Batch production	Other	None (trade, logistics...)
Scope of the model	Production / logistics network	Factory / Terminal	Manufacturing area	Plant / line / process	Component / individual consumer

Fig. 1.2 (continued)

If an in-depth investigation of the power consumption over time is required (e.g., to investigate or reduce power consumption peaks), more-detailed modeling approaches have to be selected. On the discrete side, some authors use prerecorded time series data to model energy consumption within machine states at a high resolution (Römer 2021; Römer and Straßburger 2019). On the continuous side, one may choose to model energy consumption based on the physical or chemical properties of the depicted production step (Römer et al. 2018; Pawletta et al. 2017; Peter and Wenzel 2015). Examples where a detailed continuous model for the energy aspects is beneficial include heating and cooling processes in foundries (Peter et al. 2017, Sect. 3.4.2). A third modeling alternative that has gained attention in recent research is the prediction of energy consumption using ML and AI methods (Wörrlein and Straßburger 2020a, b, Sect. 7.6.1).

Timing of the energy evaluation The second classification criterion within the dimension "Methodological Aspects of M&S" relates to the timing of the evaluation of the energy aspects. Independent from the architectural approach, the evaluation of the energy aspects can either take place concurrently to the execution of the simulation model or it can be performed in a retrospective fashion, i.e., after the simulation model has been executed.

The most-obvious choice for a concurrent evaluation is the integration of the energy aspects into a unified simulation model of the production or logistics system that includes those aspects. Then, the simulation results would typically include key performance indicators for the energy aspects, such as the total energy consumption. Other forms for a concurrent evaluation are introduced in the discussion of the criterion "Architectural approach".

A retrospective evaluation is typically based on classical outputs of the simulation model. This output enables application of evaluation measures (e.g., cost functions) to convert the usual key performance indicators (KPIs) into energy-related KPIs. Examples such as Heilala et al. (2008) are often found in the system design phase, where an environmental impact calculation including energy aspects is executed in a spreadsheet environment that exploits the results of simulation runs to evaluate certain system parameters.

Architectural approach The criterion "Architectural approach" describes how the modeling and evaluation of the energy aspects is combined with the simulation model of the production and logistics system from a computer architectural point of view. A single model, i.e., an *energy evaluation integrated in the simulation model* of the production and logistics system, may seem an obvious solution. Advantages of this approach include the centralization of the model building and the ease of model execution. It also qualifies as a concurrent approach and, thus, a further advantage is that the energy-related KPIs are directly available after the simulation run, providing for easy experimentation with the model.

Examples of this approach include simulation models based entirely on one (typically discrete) modeling paradigm (Solding and Petku 2005; Dietmair and Verl 2010) as well as hybrid simulation models based on a single simulation system supporting multiple modeling approaches (Römer and Strassburger 2019, Pawletta et al. 2017).

In some situations, it might be impossible or disadvantageous to include the energy evaluation in the simulation model itself. Reasons could lie in the desire for complexity reduction (Henriksen 2008) or the insufficient suitability of the intended tools for a specific purpose. In certain cases, the online *coupling of an evaluation tool with the simulation tool* is the architectural approach of choice (Hesselbach et al. 2008).

Such a concurrent execution of the evaluation tool and the simulation tool might not always be required. As it also imposes a certain architectural complexity concerning the execution and handling of both tools, a simpler version of coupling the simulation tools with a suitable evaluation tool is the *retrospective evaluation in a suitable tool* (Heilala et al. 2008; Johansson et al. 2009).

Another architectural approach for complementing the simulation model of the production and logistics system with an energy evaluation is the use of *separate simulation models* for the energy evaluation. Reasons may include the suitability or unsuitability of a certain simulation tool for a certain modeling task. The models can be coupled either *online* or *offline*. In the first case, they are executed concurrently in some form of distributed simulation (Peter and Wenzel 2015; Junge 2007). In the latter case they are executed sequentially.

1.2.3 Dimension 2: Objectives and Result Data

The *objective for simulation of energy aspects* is often decisive to determine the required level of detail for modeling the energy aspects and the result data to be collected. A typical objective can relate to the dimensioning and design of the production or logistics system and its energy infrastructure. In addition, a typical objective can be the optimization of energy consumption and costs. Other objectives include a load peak avoidance or reduction (Sects. 7.6.1 and 7.6.2), and, more generically, a valid forecast of energy demand.

Depending on these objectives, different types and resolutions of *relevant result data* will be required. Result data categories include KPIs directly quantifying energy aspects, such as energy consumption, energy costs, and power requirements, but also energy-related KPIs relating energy aspects of the system to other system output, such as the output quantity or the level of supply. Further result data can quantify emissions and environmental impacts. A comprehensive discussion of key performance indicators based on result data is given in Sect. 1.4.

1.2.4 Dimension 3: Input Information for Energy Aspects

The inclusion of energy-related aspects requires additional input data compared to traditional simulation models for production and logistics. First of all, additional input data are required to actually model the energy consumption and energy requirements of the equipment within the production or logistics system. These data can be based on nominal performance data of the equipment (e.g., as provided by the equipment manufacturer) or on data obtained from measurements performed on the real equipment. Different resolutions of the measured data may be useful depending on the study objective. In some scenarios, it might also be beneficial to use physical behavior models describing the energy consumption, e.g., based on differential equations. This can especially be useful if process control parameters that are subject of the simulation study influence the energetic behavior of the equipment (e.g., a heating rate of a furnace).

Secondly, additional input data may be needed for a closer evaluation of the energy aspects. Here, data about energy sources, energy prices, and energy price models, as well as data on emission equivalents may be of interest. A comprehensive discussion of input information is given in Sect. 1.3.

1.2.5 Dimension 4: Industrial Sectors and Manufacturing Type

Different industrial sectors with varying manufacturing types can have different expectations and needs when studying energy aspects within their simulation models for production and logistics. They should, therefore, be classified accordingly.

The morphological box distinguishes between the different industrial sectors that build the basis for the chapters of Part II of this book. Manufacturing is related to the majority of reported work in this field. Nevertheless, the ASIM Working Group has purposefully dedicated a separate chapter on automotive manufacturing (Chap. 3), as the automotive sector exhibits very specific requirements and applications for modeling energy aspects. The additional industrial sectors included in Part II are transportation (Chap. 4), retail (Chap. 5), perishables (Chap. 6), and renewables (Chap. 7).

Concerning manufacturing type and manufacturing structure, ASIM has relied on traditional classifications. These can have a significant influence on depicting the energy aspects. As an example, in make-to-order production, it may prove tremendously complex to obtain accurate measurement data for the energy consumption of individual orders, whereas in series production, it could be a one-time effort to measure each production step. A further classification according to the scope of the model is useful. The range of choices runs from models of entire production or logistics networks to models of individual components.

1.3 Input Information for Depiction of Energy Aspects

The quality of any simulation's results is determined by the precision of the model and the quality of its input data. Input information comprises both information that is used in modeling and data that serve as numerical inputs for the execution of simulation runs, which are also called simulation data. When production or logistics systems are investigated, these data can be categorized into the following groups (Verein Deutscher Ingenieure 2014):

- *Technical data*: Data of factory structure, production data, material flow data, and failure data
- *Organizational data*: Work time organization, allocation of resources, and structural organization
- *System load data*: Input of orders, production data

These data primarily concern the flows of material and information (Schenk et al. 2010). However, incorporating the flow of energy in a simulation study requires additional input information. While such "energy data" could be introduced as an additional group, it is also possible to integrate the respective information into the above list. The following list serves as an example of that integrated information:

- *Technical data*: Energy-related equipment data (e.g., power rating, power profiles, topology of energy distribution systems, fuel consumption of transport means like trucks or vessels, or electricity consumption for temperature regulation in warehouses)
- *Organizational data*: Utility data (e.g., contractual restrictions with energy supplier, energy market regulations, emission rules in geographic regions, or assumptions on the electric power mix)
- *System load data*: Energy procurement data (e.g., energy prices or availability of natural or market resources)

Which data are specifically required when depicting energy aspects in a simulation study greatly depends on the scope of the analysis. Data about the energy input or output of equipment will be a requirement in most cases, albeit in different form. Hence, they are usually a prime focus, especially when DES for manufacturing and logistics use cases are enhanced with energy-related aspects. Various approaches to modeling energy in DES are classified in Fig. 1.2. Likewise, a variety of different energy-related input data can be classified, where major groups are:

- *Nominal performance data (time-dependent)*: Average consumption per unit of time, e.g., in Nm^3 compressed air per hour of operation or liter diesel per driven distance
- *Nominal performance data (time-independent)*: Power rating of equipment, e.g., in kW
- *Measurement data of the real system*: Actual power demand or consumption over time, e.g., as a time series of voltage in V and current in A of an electric device over time
- *Physical behavior models*: Analytic or numeric model of the actual energy input of a device (that can typically be parameterized for specific use cases) to be used as an online or offline source of energy-related input data, e.g., a computational model of a furnace
- *Other*: Other secondary measures that relate directly or indirectly to the input of energy or emissions of relevant equipment, e.g., energy costs per hour of operation or Life Cycle Inventory (LCI) equivalent in tons of CO_2 per km driving distance

The elaborations given here primarily refer to energy input for the sake of simplicity, but can also be applied to output. Especially for technical data and depending on the applied modeling approach, a suitable data acquisition method must be selected to collect necessary input data. Table 1.1 summarizes methods that are typically employed for that purpose (cf. Kouki et al. 2017; Schmidt et al. 2015; Stoldt 2019; Weinert 2010) and their applicability for the acquisition of the above types of input data.

While this list is intended to be exhaustive, each entry classifies a variety of data acquisition processes. These follow a general concept, but can differ in the specifics, e.g., spot measurements could be applied on a work group level, on a machine level, or even on a device level. To be economical in the execution of a simulation study, it is suggested that efficiency is considered for any selection of a data acquisition

Table 1.1 Overview of energy data acquisition methods and their applicability for different types of input data (e.a. = ex ante; applicable prior to physical access to the process in question/e.p. = ex post; applicable after physical access to the process in question)

Energy data acquisition method	Short description	Other (e.g., LCI equivalents)	Nominal performance data (time-dependent)	Nominal performance data (time-independent)	Measurement data of the real system	Physical behavior models
Vendor data	Energy-related information provided by equipment vendor. Typically, such data focus on individual pieces of equipment rather than complex (sub-)systems	e.a	e.a	e.a	e.a	e.a
Life cycle inventory databases	Information usually used for life cycle assessments (LCA) provided as part of open or proprietary databases. Typically, the environmental impact of technical processes, materials, or products is quantified	e.a				
Analytic/physical model	Mathematically formulated models used for determining the energy input and output of specific physical or technical processes. Typically, only individual process steps are modeled while peripherals (e.g., consumption of control devices) are disregarded	e.a./e.p	e.a./e.p	e.a./e.p		e.a./e.p
Numeric simulation	Numeric and usually continuous models used for determining the actual or approximate energy input and output of individual processes or entire systems. Typically, implementation and validation of such models is complex and time consuming					e.a./e.p

(continued)

Table 1.1 (continued)

Energy data acquisition method	Short description	Other (e.g., LCI equivalents)	Nominal performance data (time-dependent)	Nominal performance data (time-independent)	Measurement data of the real system	Physical behavior models
Historic measurements	Existing measurement data is reused to determine energy-related simulation data. Typically, assumptions on energy input/ output of system elements or subsystems are made to quantify simulation parameters	e.p			e.p	
Spot measurements	Temporary measurements are made to determine energy-related simulation data. Typically, additional information (e.g., operation schedules) is required to quantify simulation parameters				e.p	
Continuous measurements	Continuous measurements are implemented to quantify and perpetually update energy-related simulation data. Typically, additional information (e.g., operation schedules) and complex energy data management software are required to quantify simulation parameters				e.p	
Historic consumption data	Existing metering or billing data are used to quantify typical energy input and output of individual processes. Typically, only rough estimates can be generated for the scope of the meter or bill	e.p	e.p			

(continued)

Table 1.1 (continued)

Energy data acquisition method	Short description	Other (e.g., LCI equivalents)	Nominal performance data (time-dependent)	Nominal performance data (time-independent)	Measurement data of the real system	Physical behavior models
Apportionment/ imputed appraisal	Information procured through one of the above methods for a larger (sub-)system is used to estimate simulation parameters for individual elements or subsystems. Typically requires assumptions on correct split between system components that are difficult to verify	e.p	e.p			
Comparison and estimation	Simulation parameters are estimated by comparing an object of interest with a known object for which the energy input or output are already known. Typically, a lot of know-how is required and the results will still be imprecise	e.a	e.a	e.a		

method. Efficiency is generally understood as an input–output ratio. The input here comprises all efforts that contribute to realizing the data acquisition. They can be attributed to planning and preparation, actual data acquisition, as well as additional costs (e.g., for new measurement equipment). The output comprises factors such as general applicability of a method, expected quality of the data to be generated, resolution of the resulting data, and immediacy of the application (i.e., how quickly the method can be applied). In order to identify and select the most-efficient method in a simulation study, the following heuristic process can be applied (Süße et al. 2017):

1. Analyze problem and define target for the data acquisition: Definition of required information, ex ante or ex post simulation, and level of detail as well as resolution to be achieved
2. Identify and choose alternatives: Preliminary selection of potentially suitable methods and data acquisition processes
3. Decide on decision criteria: Definition of qualitative and quantifiable criteria to measure the necessary input and the achievable output
4. Determine environmental situation: Identification of surrounding influences that might affect individual or all criteria
5. Overall assessment of alternatives and decision-making: Valuation of criteria and determination of highest efficiency, as well as final decision

While the first four steps of the suggested heuristics have arguably little structure and require both know-how and creativity from the project team, Step 5 can be supported by more-structured decision-making tools. Specifically, the *analytic hierarchy process* or *utility value analysis* methods (Götze et al. 2015) can be used to support the valuation.

A guiding principle according to VDI 4633 Part 1 states that "Modeling accuracy should not be as detailed as possible, but as detailed as necessary to fulfil the given targets" (Verein Deutscher Ingenieure 2014, p. 6). This is of particular importance when energy aspects are modeled in addition to the flow of materials. Overall, the target measures to be quantified in the simulation need to be appropriate to the scope of the simulation study, the modeling must be suitable to serve the measures, and the data acquisition must be sufficient for the modeling requirements. As such, it is impractical to collect needlessly precise and high-resolution data on the energy input or output of system elements, when only a very highly aggregated consumption measure is sought. If high-resolution data or physical or numerical simulation models are reused or also collected for other purposes, it may be necessary to apply preprocessing that reduces the quantity of data provided to the simulation.

Nevertheless, there are no general rules that suggest a resolution for input data based on the measures to be quantified through a simulation. Studies on machine tools have demonstrated, however, that machines with different technologies exhibit substantially different load cycles and distributions of demand between individual machine parts (Wegener and Weiss 2014). Accordingly, it is suggested to take these aspects as well as the scope of the investigation into account when deciding on the resolution of input data to be acquired. The scope is best considered when asking the question: What is the economic or technical aspect that is focused on in the study? For instance, billing by utility companies takes the overall electric work in Wh as well as the 15-min maximum power demand in W into account. If billing-relevant measures are within the scope and the maximum power demand can be assumed to be of little interest, due to low resolution data, only the margin of error must be considered when deciding on the resolution of input data. On the contrary, if technical questions regarding the suitability of an unbuffered energy generator are considered, it is of great importance to know exactly when multiple peaks overlap within the system. For such studies, a high-resolution and precise knowledge on the uniformness of patterns as well as on sources for disturbances in the energy input and output behavior are required.

Just like other simulation data, energy-related input information requires verification and validation (V&V) to ensure that the simulation results are sufficiently accurate with respect to the (future) real-world system. The V&V methods to be applied for that purpose are the same as for simulation that disregards the flows of energy, including verification techniques such as extreme condition tests and sensitivity analyses as well as validation approaches like face validity and touring tests (cf. Banks et al. 2005; Rabe et al. 2008). Typically, the most-difficult aspect to validate is the energy input or output of the system over time. Experience shows that energy-related input data should be validated on the lowest possible level rather than the system level to avoid the possibility that errors in the input data of individual elements mutually compensate in a specific reference scenario, but stack when other circumstances are simulated. Ideally, this validation has already been conducted on the input data provided to the simulation, e.g., by comparing the energy consumption profile to be used with multiple instances of the same cycle measured from real equipment. This comparison can be conducted in a qualitative manner using graphs or by performing a regression analysis. When no such possibility exists, because, for instance, the system to be simulated has not yet been implemented, it is even more important to estimate the margin of error to be expected for each system element. This can help to assess the modeled system's overall margin of error.

1.4 Key Performance Indicators

The identification and the respective selection of key performance indicators (KPIs) is a crucial step that has to precede almost any planning task and most certainly any simulation model development (Verein Deutscher Ingenieure 2014). Typically, abstract and highly aggregated targets need to be broken down into measurable or at least more easily assessable targets. An example would be the analysis of different production control strategies with the overall objective to maximize profitability, which could be broken down into the minimization of throughput times, inventory, and schedule deviations, as well as the maximization of capacity utilization (Verein Deutscher Ingenieure 2014). Depending on the actual context of an investigation or planning activity, even-more-technical considerations and KPIs become important, when energy is considered as a prominent aspect. For instance, the dimensioning of a compressed air system requires specific information on the consumption of different consumers, enabling the calculation of demand coverage to assess the viability of a proposed supply setup. In light of these considerations, it is apparent that practically relevant energy-related KPIs can be manifold. The categories of such KPIs and corresponding result data that are typically relevant in production and logistics applications have been shown in Table 1.1. These are further explained in Table 1.2.

These categories can be characterized based on their scope (flow-item-related or system-related) and their relation to time (Fig. 1.3). It is apparent from these characterizations that the result data of simulation studies can differ significantly from analytic calculations. At the same time, there are also significant overlaps. For the purpose of this chapter it is assumed that the reader has a solid understanding of simulation worthiness (Verein Deutscher Ingenieure 2014). Therefore, the remainder of the elaborations will omit a discussion of when simulation would not be an appropriate tool and instead will focus on aspects where it can serve to provide additional insights.

In contrast to static or analytic methods, simulation provides capabilities to effectively model and analyze the dynamics of systems concerning their energy consumption or emission behavior. As such, it can provide greater precision insights in the following ways:

1. **Dynamic influences among elements** Static approaches can require the definition of certain load or requirement scenarios, from which conclusions on the respective demand or emission behavior can be drawn. Dynamic influences between elements are only considered using empirically obtained concurrency factors or similar measures along with possibly substantial security factors. While this can be a feasible and a valid procedure in many cases, it is primarily masking missing knowledge about the system's specifics. Simulation can contribute to transparency in such situations, because the actual system behavior (albeit potentially abstracted) is modeled. Hence, concurrency scenarios among system elements can be created within the simulation and conclusions drawn from these. Such considerations will naturally require the application of time-dependent KPIs that are mostly focused on the system and its elements.
2. **Stochastic influences of elements** Besides dynamic influences, stochastic influences are a primary source of imprecision for static approaches to the dimensioning of energy systems. Accordingly, they are also mostly respected using concurrency and safety factors. Simulation, on the other hand, can be used to vary the influence of stochastic properties in a structured manner. Thus, it is possible to determine with greater confidence expectable extreme scenarios that must be considered when considering time-dependent KPIs. Furthermore, the calculation of time-independent KPIs can benefit from the simulation of stochastic influences and establish a more-reliable level of confidence.
3. **Time-dependence of constraints** Simulation also provides significant benefits when constraints, which must be observed in the calculation of KPIs, are time-dependent themselves. For instance, when energy prices or energy supply change

Table 1.2 KPI and result data categories for energy-related simulation studies

Category	Description	Exemplary KPIs
Energy consumption	KPIs that measure the consumption of energy on an arbitrary level of the system hierarchy. Depending on the scope of the study, only a single summarizing value or time series data are sought. They are usually focused on a single energy carrier (e.g., only to electricity)	System electricity consumption in kWh Compressed air consumption within 5-min intervals over time in Nm^3/min Energy consumption per product in kWh/piece
Energy costs	KPIs that quantify the economic impact of energy consumption. While generally determined from the energy consumption, the underlying pricing model can vary depending on the scope of the study. Costs can be aggregated over multiple energy carriers (e.g., gas and electricity)	Natural gas costs in € Overall electricity costs in €/h operation Energy costs per product in €/piece
Power requirement	KPIs that can be used to assess the power requirements of energy infrastructure systems. They are generally time-dependent, but can be broken down to specific aspects of the time series data (e.g., peak load). The focus is only on a single energy carrier	Peak electricity load in kW Average off-shift load in kW Electric power demand over time in kW
Emissions and environmental impacts	KPIs that measure the emissions and environmental impacts of a system directly or indirectly. They are similar to energy consumption KPIs, but typically rely on equivalence data (e.g., kgCO_2/km) and could be determined indirectly (e.g., from electricity consumption). Typically, no time series data are sought	NO_2 emissions in t NO_2 Emitted CO_2 equivalent in t CO_2 Primary energy demand per product in MJ/piece
Energy-related key performance indicators	KPIs that quantify specific aspects of the energy conversion and transmission systems. Exemplary aspects are level of supply, self-sufficiency ratio, energy output (e.g., as heat) of individual system elements, load balance, etc. They are focused on a single-energy carrier	Self-sufficiency ratio for electricity system (e.g., with photovoltaic self-supply) in % Waste heat in J Utilization ratio of air compressor in %
Other	KPIs that allow for the qualitative or semi-quantitative assessment of energy-related aspects through surrogates. Surrogates have a proven but not quantified relationship to the consumption of energy	Time shares for operation states in % Value added ratio of energy-consumption in %

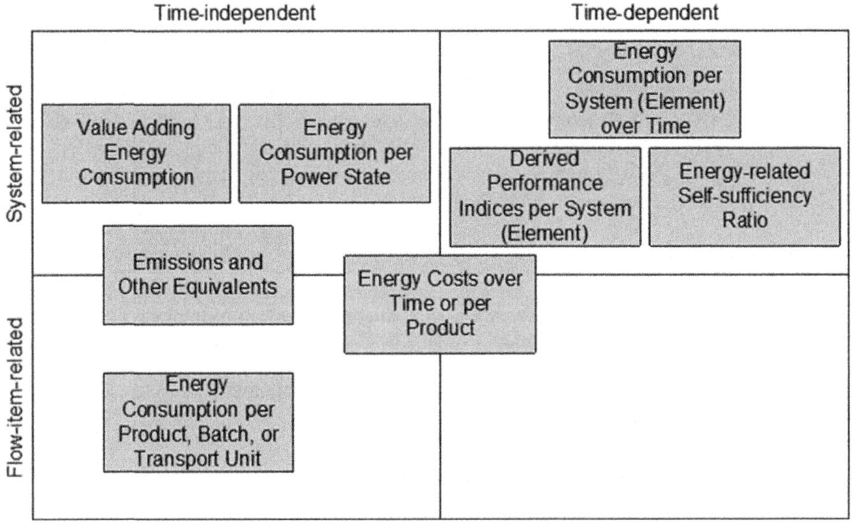

Fig. 1.3 KPIs for energy-related aspects (adapted from Uhlig et al. 2018, 3281)

over time, it becomes increasingly complex to correctly include their respective changes in a static or analytic calculation. Their effects can be attributed more easily to both time-dependent and time-independent KPIs with a system-related or a flow-item-related scope utilizing simulation.

4. **Time series analysis** While static methods allow for calculating results for different scenarios, they generally are not well-suited to aggregate the behavior of multiple individually operating elements over the course of time. In contrast, simulation provides the means to collect and export results as time series data. On this basis, other methods (e.g., time series analysis, machine learning, regression analysis, etc.) can be applied to create additional insights on the subject matter. Such post-processing is not necessarily required to obtain knowledge on certain aspects of time series data from a simulation. Alternatively, triggers can be implemented into a model to collect, for instance, maximum peak load data during simulation runtime.
5. **Scenario building** The utilization of static and analytic methods can be facilitated with a variety of means, e.g., pen and paper, spreadsheet software, or computer algebra system (CAS) software. However, even when software is used, it requires the executing engineer to follow certain design patterns that allow for the execution of scenario experiments later. The general process of simulation studies as well as the setup of modern simulation tools ensures that experimentation can always be facilitated with reasonable ease. Thus, KPIs can be recalculated for different and possibly incrementally defined scenarios.

A primary prerequisite to attain a benefit from using simulation in the calculation of KPIs is that the modeling actually adds an aspect to the calculation that would otherwise not be assessable. For instance, if simulation was used to quantify the energy costs of a system that is modeled without stochastic, dynamic, or time-dependent influences, this could be done using static or analytic means. On the other hand, if the energy prices vary over time or are also dependent on the peak load within 15-min intervals, this can be difficult—if at all possible—to assess analytically or with static spreadsheet calculations.

The above elaborations have already referenced various energy carriers that are used in production and logistics. From a physical point of view, the most-prevalent types of energy being used in these sectors are chemical energy (e.g., fuel), electric energy (electricity), kinetic energy (e.g., compressed air), and thermal energy (e.g., coolant liquid). KPIs can focus on any of these and may also cover multiple stages of energy conversion, e.g., overall electricity consumption that takes the consumption of a compressed air system along with the direct consumption of electricity of all system elements into account. Whether such multi-stage considerations are sensible or whether all energy carriers in a system need to be observed strongly depends on the goals of the task at hand. Furthermore, sometimes they can be abstracted altogether, especially when KPIs on emissions, emission equivalents, or environmental impacts are sought.

The consideration of costs provides additional challenges that go beyond the modeling of the physical flows of energy. Pricing can follow several constraints that must be observed with great care, prompting for specific in-depth knowledge on the price-making mechanisms during the modeling phase. For instance, the overall electricity costs for industrial companies often split into two components: (1) consumption-based pricing, and (2) load-based pricing. The former is the product of consumed electricity (i.e., electric work) and agreed price per unit of measurement. The latter is determined through the highest load during a 15-min interval throughout the billing period. Further intricacies stem from the pricing of certain commodities, such as natural gas. They can be of great importance when long periods of time are being analyzed. Lastly, energy costs for energy carriers that are locally converted (e.g., compressed air), but sometimes also those which are directly sourced, actually comprise fixed as well as variable costs. Whether both aspects are included in the KPI calculation must be determined based on the specific task. From an accounting perspective, fixed costs and their allocation over time or system elements can be delicate. Therefore, special care must be taken during the modeling phase.

The trend towards the increasing integration of renewable energy sources into the power grids but also for on-site energy generation has broadened the scope for energy-related KPIs in production and logistics. On the one hand, new KPIs, such as the self-sufficiency in electricity supply, have become of interest for simulation. On the other hand, established KPIs were repurposed (e.g., to quantify system loads for thermal cooling or heating for the dimensioning of geothermal facilities) or required more extensive calculations for their quantification (e.g., when CO_2 emissions are compensated through the partial generation of renewable electricity).

References

Banks J, Carson JS II, Nelson BL, Nicol DM (2005) Discrete-event system simulation, 4th edn. Prentice Hall, Pearson

Colquhoun HL, Levac D, O'Brien KK, Straus S, Tricco AC, Perrier L, Kastner M, Moher D (2014) Scoping reviews: time for clarity in definition, methods, and reporting. J Clin Epidemiol 67(12):1291–1294

Dietmair A, Verl A (2010) Energy consumption assessment and optimisation in the design and use phase of machine tools. In: Proceedings of the 17th CIRP international conference on life cycle engineering, Hefei, China, May, pp 116–122

Götze U, Northcott D, Schuster P (2015) Investment appraisal: methods and models, 2nd edn. Springer, Berlin Heidelberg

Heilala J, Vatanen S, Tonteri H, Montonen J, Lind S, Johansson B, Stahre J (2008) Simulation-based sustainable manufacturing system design. In: Mason SJ, Hill RR, Mönch L, Rose O, Jefferson T, Fowler JW (eds) Proceedings of the 2008 Winter Simulation Conference. IEEE, Piscataway, NJ, pp 1922–1930

Henriksen JO (2008) Taming the complexity dragon. J Simul 2(1):3–17

Hesselbach J, Herrmann C, Detzer R, Martin L, Thiede S, Lüdemann B (2008) Energy efficiency through optimized coordination of production and technical building services. In: Proceedings of the 15th CIRP international conference on life cycle engineering. Sydney, Australia, pp 624–629

Johansson B, Kacker R, Kessel R, McLean C, Sriram R (2009) Utilizing combinatorial testing on discrete event simulation models for sustainable manufacturing. In: Proceedings of 2009 ASME design for manufacturing and the life cycle conference. San Diego, California, USA, pp 1095–1101

Junge M (2007) Simulationsgestützte Entwicklung und Optimierung einer energieeffizienten Produktionssteuerung. Ph.D. thesis, University Kassel, Produktion & Energie, Kassel

Kouki M, Cardin O, Castagna P, Cornardeau C (2017) Input data management for energy related discrete event simulation modelling. J Clean Prod 141:194–207

Pawletta T, Schmidt A, Junglas P (2017) A multimodeling approach for the simulation of energy consumption in manufacturing. Simul Notes Europe 27(2):115–124

Peter T, Wenzel S, Reiche L, Fehlbier M (2017) Coupled simulation of energy and material flow—a use case in an aluminum foundry. In: Chan WKV, D'Ambrogio A, Zacharewicz G, Mustafee N, Wainer G, Page E (eds) Proceedings of the 2017 Winter Simulation Conference. IEEE, Piscataway, NJ, pp 3792–3803

Peter T, Wenzel S (2015) Simulationsgestützte Planung und Bewertung der Energieeffizienz für Produktionssysteme in der Automobilindustrie. In: Rabe M, Clausen U (eds) Simulation in production und logistics 2015. Fraunhofer, Stuttgart, pp 535–544

Poeting M, Prell B, Rabe M, Uhlig T, Wenzel S (2019) Considering energy-related factors in the simulation of logistics systems. In: Mustafee N, Bae KHG, Lazarova-Molnar S, Rabe M, Szabo C, Haas P, Son YJ JW (eds) Proceedings of the 2019 Winter Simulation Conference. IEEE, Piscataway, NJ, pp 1849–1858

Rabe M, Spieckermann S, Wenzel S (2008) Verifikation und Validierung für die Simulation in Produktion und Logistik. Springer, Berlin Heidelberg

Römer A, Strassburger S (2019) Hybrid system modeling approach for the depiction of the energy consumption in production simulations. In: Mustafee N, Bae K-HG, Lazarova-Molnar S, Rabe M, Szabo C, Haas P, Son YJ (eds) Proceedings of the 2019 Winter Simulation Conference. IEEE, Picataway, NJ, pp 1366–1377

Römer A, Rückbrod M, Straßburger S (2018) Eignung kombinierter Simulation zur Darstellung energetischer Aspekte in der Produktionssimulation. In: ASIM 24th Symposium Simulationstechnik. Hamburg, Germany, pp 73–80

Römer A (2021) Simulation-based optimization of energy efficiency in production. Springer Gabler, Wiebaden

Schenk M, Wirth S, Müller E (2010) Factory planning manual. Springer, Berlin Heidelberg

Schmidt C, Li W, Thiede S, Kara S, Herrmann C (2015) A methodology for customized prediction of energy consumption in manufacturing industries. Int J Precision Eng Manufact-Green Technol 2:163–172
Solding P, Petku D (2005) Applying energy aspects on simulation of energy intensive production systems. In: Kuhl ME, Steiger NM, Armstrong FB, Joines JA (eds) Proceedings of the 2005 Winter Simulation Conference. IEEE, Piscataway, NJ, pp 1428–1432
Stoldt J, Schlegel A, Putz M (2016) Enhanced integration of energy-related considerations in discrete event simulation for manufacturing applications. J Simul 2(10):113–122
Stoldt J, Prell B, Rabe M, Wenzel S, Thiede S (2021) A Criteria-based database for research and applications of energy-oriented simulation in production and logistics. In: Franke J, Schuderer P (eds) Simulation in Produktion und Logistik 2021. Cuvillier, Göttingen, pp 93–102
Stoldt J (2019) Gestaltungsmethodik für Simulationsstudien in Umplanungsprojekten zur Energieeffizienzsteigerung in Fabriken. Ph.D. thesis, TU Chemnitz, Mechanical Engineering, Chemnitz
Sturma A, Ritschl V, Dennhardt S, Stamm T (2016) Reviews. In: Ritschl V, Weigl R, Stamm T (eds) Wissenschaftliches Arbeiten und Schreiben. Springer, Berlin Heidelberg, pp 207–221
Süße M, Stoldt J, Schlegel A, Putz M (2017) Decision procedure for efficient data acquisition for distinct energy prognosis models. Appl Mech Mater 871:114–124
Uhlig T, Wenzel S, Peter T, Stoldt J, Schlegel A, Jósvai J (2018) Considering energy in the simulation of manufacturing systems. In: Rabe M, Juan A, Mustafee N, Skoogh A, Jain S. Johansson B (eds) Proceedings of the 2018 Winter Simulation Conference. IEEE, Piscataway, NJ, pp 3275–3286
Verein Deutscher Ingenieure (2014) VDI Guideline 3633 simulation of systems in materials handling, logistics and production—Part 1. Fundamentals. Beuth, Berlin
Wegener K, Weiss L (2014) Auslegung von ressourceneffizienten Produktionsprozessen und Maschinen. In: Proceedings 3rd International Chemnitz manufacturing colloquium, Chemnitz, 8th–9th April, part 1, pp 61–82
Weinert N (2010) Vorgehensweise für Planung und Betrieb energieeffizienter Produktionssysteme. Fraunhofer, Stuttgart
Wenzel S, Peter T, Stoldt J, Schlegel A, Groß G, Pitsch H, Rabe M, Seewaldt M (2017) Betrachtungen energetischer Einflussfaktoren in der Simulation in Produktion und Logistik: eine Literaturanalyse. In: Wenzel S, Peter T (eds) Simulation in Produktion und Logistik 2017. Kassel university press, Kassel, pp 9–18
Wörrlein B, Straßburger S (2020a) A method for predicting high-resolution time series using sequence-to-sequence models. In: Bae KH, Feng B, Kim S, Lazarova-Molnar S, Zheng Z, Roeder T, Thiesing R (eds) Proceedings of the 2020 Winter Simulation Conference. IEEE, Piscataway, NJ, pp 1075–1086
Wörrlein B, Straßburger S (2020b) On the usage of deep learning for modelling energy consumption in simulation models. Simulation Notes Europe 4(30):165–174

Part II
Application Fields

Chapter 2
Manufacturing

Sebastian Thiede, Antal Dér, Marc Münnich, and Thomas Sobottka

Abstract The manufacturing industry is responsible for a large share of global environmental impacts (e.g., greenhouse gas emissions) that can mainly be tracked back to energy demand. This energy demand is determined by a diversity of processes and machines, which dynamically interact in process chains and with other factory elements such as technical building services (TBS). Given that, system-oriented material flow simulation with inclusion of energy aspects bears the potential to support the energy transition of industry through fostering both energy efficiency and substitution towards renewable resources. The chapter addresses the necessary background as well as common aspects in the context of energy-oriented manufacturing system simulation. Four manufacturing case studies underline the feasibility and potential of available simulation approaches for improving energy-related environmental impacts and also costs. Additionally, an outlook towards potential future research steps is given.

S. Thiede (✉)
Chair of Manufacturing Systems, Department of Design, Production and Management, Faculty of Engineering Technology, University of Twente, Drienerlolaan 5, Enschede 7522NB, The Netherlands
e-mail: s.thiede@utwente.nl

A. Dér
Sustainable Manufacturing and Life Cycle Engineering, Institute of Machine Tools and Production Technology (IWF), Technische Universität Braunschweig, Langer Kamp 19B, 38106 Braunschweig, Germany
e-mail: a.der@tu-braunschweig.de

M. Münnich
Fraunhofer Institute for Machine Tools and Forming Technology IWU, Reichenhainer Straße 88, 09126 Chemnitz, Germany
e-mail: marc.muennich@iwu.fraunhofer.de

T. Sobottka
Fraunhofer Austria Research GmbH, Theresianumgasse 7, 1040 Vienna, Austria
e-mail: thomas.sobottka@fraunhofer.at

S. Wenzel et al. (eds.), *Energy-Related Material Flow Simulation in Production and Logistics*, https://doi.org/10.1007/978-3-031-34218-9_2

2.1 Introduction

Industry is responsible for a major share of human-induced global environmental impacts—as just one example, the industrial sector counts up to around a third of global greenhouse gas emissions. This includes direct and indirect emissions along the whole manufacturing value chain with both process industry and discrete manufacturing industry (IPCC 2014; Thiede 2021). The major share of these emissions is caused by the energy demand of production processes and the factories as a whole.

The remainder of this chapter is organized as follows: Sect. 2.2 contains an overview of the scope and objectives for the simulation of energy aspects in manufacturing systems in general. In Sect. 2.3, the simulation approaches in this application field are discussed. The following four sections illustrate these with different kinds of related examples. The chapter ends with a short conclusion and outlook for further developments (Sect. 2.5).

2.2 Scope and Objectives

Figure 2.1 exemplary shows the composition of the carbon footprint of an automotive factory over a period of 30 years (Gebler et al. 2020). Quite clearly, the energy demand of the use phase is dominating here, followed by the embodied emissions of the technical equipment and necessary auxiliary materials to run the factory. But, these emissions are also caused by energy demand that was needed for producing this equipment and material upstream at their respective suppliers. The figure also clearly underlines the necessity of a holistic factory system understanding: A factory consists of three main elements—the production equipment (e.g., milling or turning machines), technical building services (TBS, e.g., HVAC—heating, ventilation, air conditioning, energy supply) and the building shell (Thiede 2012, Fig. 2.2). The design and control of those elements and their interaction eventually determines the energy demand of the factory. Related to the changing operational states of all elements, the total energy demand is not static but very dynamic. Figure 2.3 shows a typical resulting load profile, which represents the total energy demand over time (Dehning et al. 2019). Characteristic indicators like base load or occurring peaks are of interest for companies for analysing, benchmarking, reporting, and improving energy demand.

In order to achieve a reduction of energy-induced environmental impact in manufacturing, two main strategies can be distinguished (Table 2.1, e.g., described in Thiede 2021). Energy efficiency aims at improving the ratio of production output (products) and the necessary energy input. Therewith, the specific energy demand per product would be decreased. In contrast to that, energy flexibility rather addresses the effectiveness strategy, e.g., aiming at substitution of energy carriers to less costly or renewable alternatives. This can be facilitated through synchronizing the manufacturing energy demand with the current energy supply. Therewith, high energy

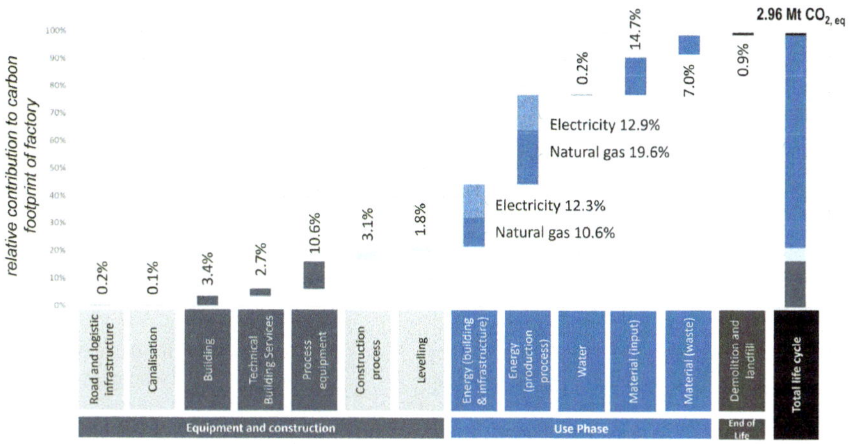

Fig. 2.1 Carbon Footprint of an automotive factory (adapted from Gebler et al. 2020)

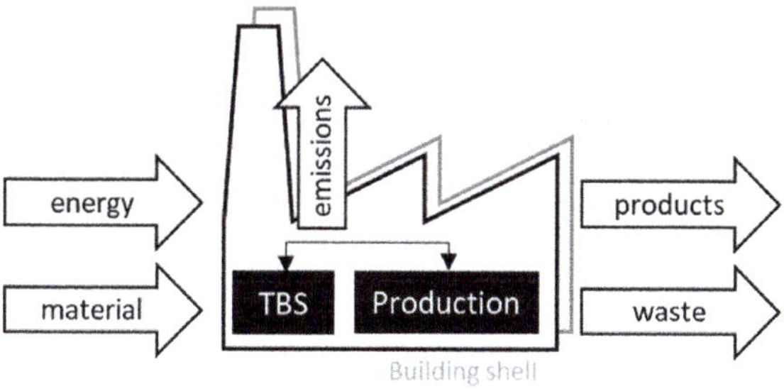

Fig. 2.2 Overview holistic factory system understanding (adapted from Thiede 2021) (TBS: technical building services)

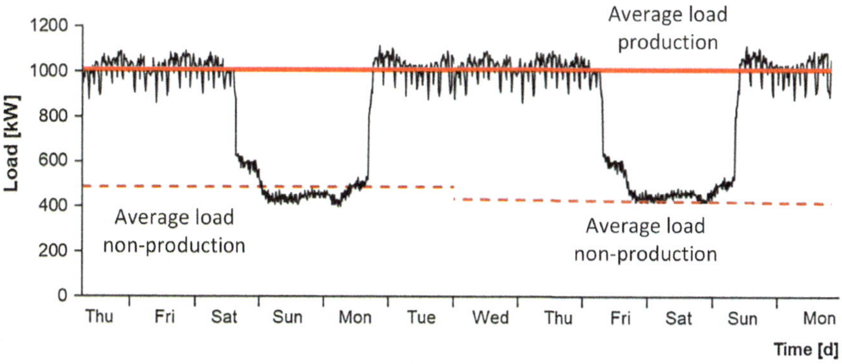

Fig. 2.3 Exemplary energy load profile of an automotive factory (adapted from Dehning et al. 2019)

Table 2.1 Fields of action for improving energy-induced environmental impact in factories

	Energy efficiency	Energy flexibility
Idea	Improving the ratio of valuable output and energy input in manufacturing	Time-based alignment of manufacturing energy demand to energy supply
Potential sustainability impacts	*Efficiency* strategy with stronger decoupling of output and input leading to potential savings while keeping the output	*Effectiveness* strategy while shifting (potentially higher) energy demand to times of lower costs or lower environmental impact. Potential energy autarky with renewable energy sources
Enabling technologies	Technical measures for improving the efficiency of technical systems, e.g., more energy-efficient drives and low idle energy demand	Technical systems for the necessary flexibility potential, energy storage typically needed to utilize full potential
Methods and tools	Data-based modeling and simulation for energy data analysis, identification of most-relevant subsystems and influencing factors as well as derivation and virtual testing of improvement actions	

demand shall be shifted to time slots with lower costs or less environmental impact, e.g., when a high share of renewable energy sources is available in the electricity grid or even through on-site generation. This can play an important role for improving the environmental impact and energy autarky of the specific company. But, given the energetic relevance of manufacturing industry, this is also relevant for the overall transition of energy grids. Both technical and organizational measures are possible to increase energy efficiency or flexibility (many examples can be found in, e.g., Sauer et al. 2019). However, strong interdependencies are given among all factory elements and towards other key performance indicators in manufacturing systems.

2.3 Energy-Oriented Simulation Approaches for Manufacturing Systems

As indicated, considering dynamic interdependencies in the whole manufacturing system is important when analysing and improving energy demand. Even more, energy is of course only one perspective, while other production key performance indicators related to output quantity, time, quality, or costs need to be taken into account as well. All these aspects call for using manufacturing system simulation and a respective consideration of energy demands.

Using simulation for improving planning and operation of manufacturing systems is established for several decades. However, energy-related aspects were just introduced in the last 10–15 years. Thiede et al. (2013) gave an early overview of respective approaches. A more recent state-of-the-art analysis can be found in Walther and

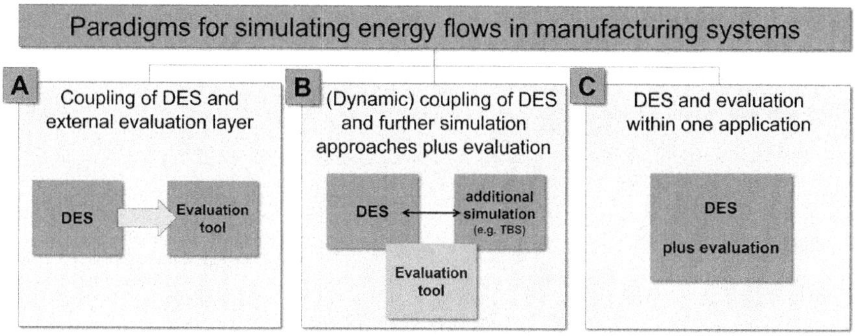

Fig. 2.4 Paradigms for energy oriented manufacturing system simulation (Thiede 2012; Herrmann et al. 2011)

Weibold (2021). In general, three paradigms for energy integration can be distinguished (Fig. 2.4). Paradigm A adds an external energy evaluation to the simulation environment (e.g., simulation engine provides production data, which will be externally used for energy demand calculation). With paradigm C, this energy consideration is directly included in the same environment. Paradigm B couples the production simulation with further modules (e.g., to integrate technical building services), either internally in one software environment or through coupling of different tools.

Another recent analysis of energy-oriented simulation publications over the years (Fig. 2.5) confirms the rise of the topic as such but also the variety of approaches and their connections to different paradigms (Stoldt et al. 2021). Figure 2.6 illustrates the deployed simulation architectures and connected simulation tools. All described paradigms can be found, but especially paradigm C has been on the rise in the last years. A major reason lies in the integration of energy-related aspects into the tool Siemens Plant Simulation, which is an established simulation software in research and industry. Therewith, entry barriers for integrating energy aspects were significantly lowered. However, when it comes to coupling of different models, also other simulation tools (e.g., AnyLogic, Matlab) or simply proprietary developments play a strong role as well, e.g., if physical energy system behavior is to be modeled or simulation-based optimization requires high simulation computing performance.

2.4 Applications

On this background, in the following sections four case studies will be presented that depict different approaches for an energy-oriented simulation of manufacturing systems. All those cases consider different energy flows to eventually improving energy-induced costs or the carbon footprint of manufacturing (or other environmental impacts), while taking general production performance objectives (e.g., time, utilization) into account as well. However, as depicted in Table 2.2, the approaches

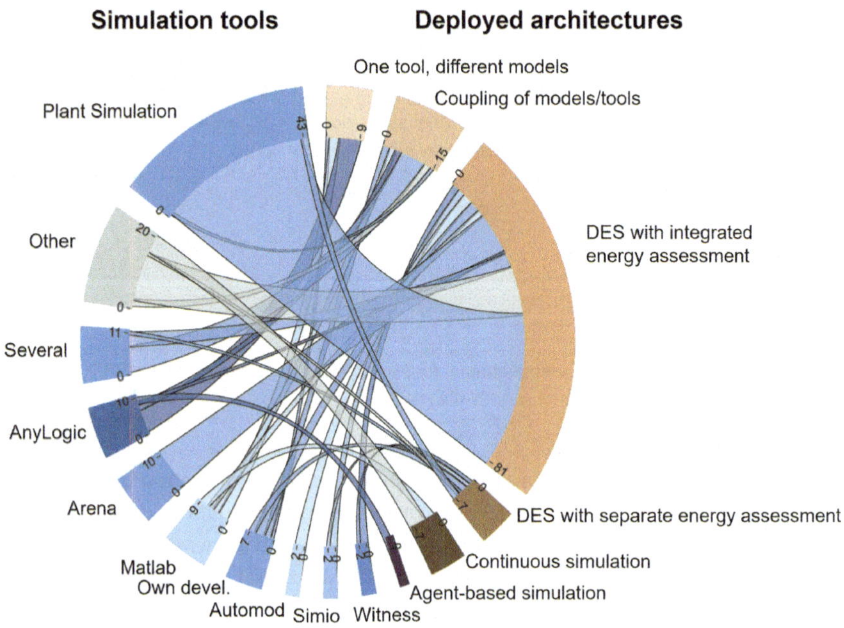

Fig. 2.5 Analysis of approaches for energy-oriented factory simulation (Stoldt et al. 2021)

differ significantly in terms of considered types of processes and factory elements as well as underlying simulation architectures and tools (Table 2.2).

2.4.1 *Heat Treatment in a Casting Company*

This case study is situated at an Austrian casting manufacturer and focusses on the simulation-based optimized planning of heat treatment processes, considering a goal function that includes energy efficiency alongside traditional economic goals (costs- or profit-driven) in production planning and control (PPC). A more-detailed description of this case can be found in Sobottka et al. (2020). Cast steel production is an energy-intensive production. Therefore, rendering energy optimization via planning and control in this industry is especially beneficial. The heat treatment furnaces and transient heat-exchange interactions with their surroundings as well as the processed workpieces constitute a complex thermodynamic behavior, which is captured simultaneously with the production logic and material flow behavior in a suitable planning method via a hybrid discrete–continuous simulation.

Table 2.2 Case study comparison

	Case study 1: Casting company	Case study 2: Machining process chains	Case study 3: Hot forging process chains	Case study 4: Battery cell factory
Application domain/focus processess	Heat treatment processes	Machining processes (milling, turning etc.)	Hot forging process chains	Battery manufacturing process chain in factory context
Energy carriers	Electricity, gas	Electricity	Electricity	Electricity, gas
Factory elements	Furnaces, energy supply system	Diverse machine tools	Diverse production machines	Production machines, compressed air, HVAC
Simulation architecture	Hybrid continuous-discrete simulation, production and energy system and evaluation/ optimization in one tool	Agent-based modeling and simulation in one tool	Discrete event simulation with integrated energy evaluation	Hybrid co-simulation with coupled models
Main tool for implementation	Free programming/ own development	Anylogic	Plant simulation	Anylogic and Matlab Simulink

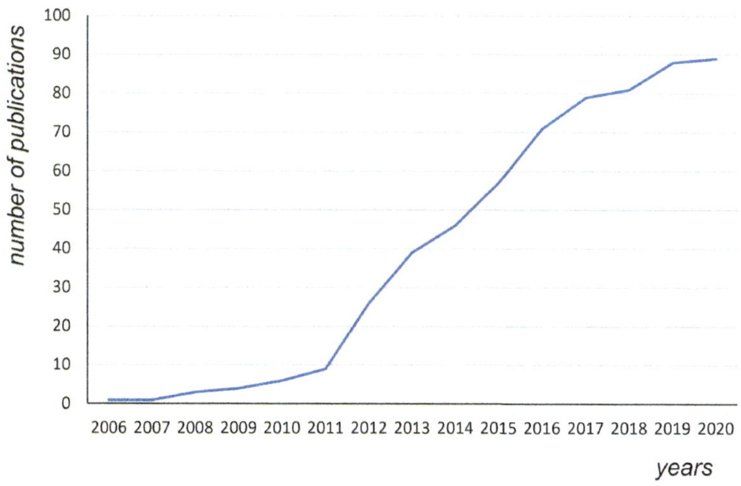

Fig. 2.6 Cumulated number of publications over time (Stoldt et al. 2021)

2.4.1.1 Objectives

The system in the case study comprises of the loading of work pieces on to grates, which are then lifted to input buffers of one out of five parallel heat treatment furnaces. Four of the furnaces run on natural gas and one is powered by electricity. Upon exiting the furnaces, the workpieces are forwarded by the same crane to different cooling stations. There are roughly 150 different heat treatment programs, consisting of nine different heat treatment processes (clearing, annealing, warm-up, relaxation, glowing, normalizing, air/water/oil-based steel tempering), all differing in their individual temperature profiles and heat treatment durations for each individual product type. Technological restrictions apply concerning which program is executable on a given furnace. The upstream casting process and its schedule are input parameters for this case study. Downstream, beyond the system boundaries, there are less energy-intensive machining operations that are less risky to form a bottleneck for the entire system. The actuating variables for the planning method are: batching of orders to furnace grates, assigning the batches to furnaces, scheduling and sequencing the processing of the batches in the furnaces, and controlling the furnaces (times for switching the machines on and off and, thus, control of the heating process).

2.4.1.2 Setup of the Simulation Approach

In the planning method, a simulation-based optimization is implemented, utilizing the hybrid simulation of the production system as an evaluation function for an optimization. An overview of the entire method is shown in Fig. 2.7. In this setup, the optimization modulates the actuating variables and initiates multiple simulation runs to evaluate the fitness of intermediate solutions via a multi-criteria objective function, until the stopping criteria – allowed optimization time or convergence of the goal fitness values—are reached (Sobottka et al. 2018).

Concerning the above-mentioned energy simulation paradigms, the simulation approach at hand is a special form of C in Fig. 2.4): the production simulation (DES) is integrated with a simulation for the energy system (a continuous simulation). However, the integration is not implemented by connecting two or more simulators in a co-simulation, but at a building block level of the simulator (Sihn et al. 2018). Each simulation building block, atomics, can create and process events as well as solve differential equations, thus directly enabling interactions between thermo-physical behavior and production-logic events. With the hybrid building blocks, complex models can be built. This also allows for an object-oriented modular simulator, for which libraries of frequently used production system elements, such as basic machine types of logistics equipment, can be developed, instantiated, and reused in new applications. This novel simulator increases the level of integration between the simulation domains—in co-simulation, integration is limited by computing effort—and it also improves the practical applicability of the method, since models or parts of models can be reused, reconfigured, and applied to new cases more efficiently

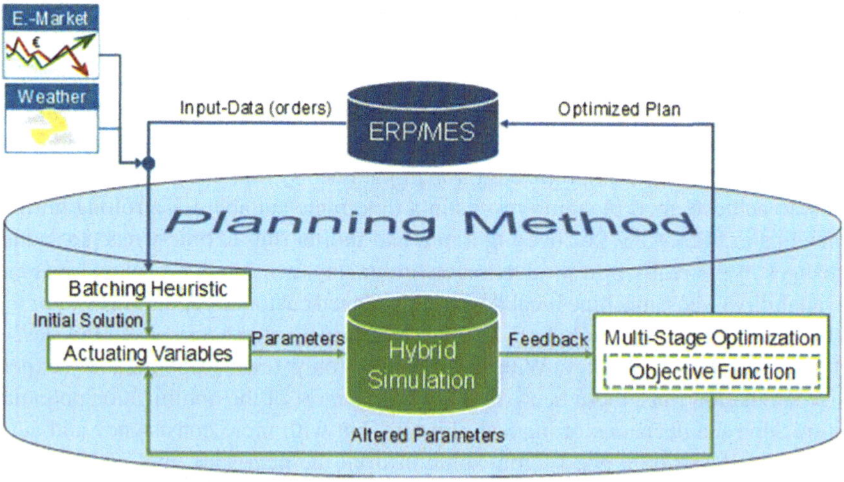

Fig. 2.7 Planning method with hybrid simulation at its core (Sobottka et al. 2020)

than with co-simulation, where re-use of model elements is usually infeasible (Sihn et al. 2018).

The developed optimization method consists of a multi-stage optimization, hybridizing rule-based heuristics with metaheuristics. The first stage of the optimization is an order batching heuristic that prioritizes the technically demanding orders – those that require a special furnace – over those that can be processed on all furnaces. It creates "crystallization points" for pressing, technically demanding orders and fills up the associated furnace grates with suitable other orders, according to their due dates and in a certain due date range.

The following second stage, a deterministic exchange-based heuristic, evaluates pairwise exchanges between similar—in duration and technological requirements—batches with the simulation as a fitness evaluation. With the given problem complexity, this heuristic can evaluate all exchange opportunities enumeratively.

Taking the results of the exchange heuristic as the initial solution, the third optimization stage, a customized genetic algorithm (GA), modulates the timeslots and sequence of batches in the furnaces, also using the simulation as its evaluation function.

2.4.1.3 Results

Experimental studies for the case study have been conducted with historical data from the real-world production. With planning horizons of two weeks, the production plans were optimized using the developed multi-stage-multi-method approach. The optimization quality was evaluated compared to the manual planning results provided by the production planners. These manual reference results were roughly equivalent to the results of the first optimization stage, the batching heuristic. Concerning the global fitness value of the objective function, the optimization was able to achieve a 10% improvement, with the energy costs improved by 6%. The associated CO_2 emissions, extrapolated to yearly savings, can be reduced by 200 tons per year. Figure 2.8 shows the fitness value trends of the objective function during the optimization. The optimization was itself optimized, through parametrization and customized operators, to achieve good planning results in a timeframe suitable for a rolling horizon planning in MES/APS, i.e., overnight runs and shorter runs to timely react to sudden changes in the real-world production environment, such as personnel and material availability issues, machine breakdowns, or changed customer requirements. For this efficient optimization the hybrid approach with rule-based heuristics at the start of the optimization is the key. With no or at least only few computationally expensive simulation runs, these heuristics can realize most of the optimization potential, mimicking the decisions of manual planners, but with more consistency and automatically. Based on a good initial solution from the heuristics, the metaheuristics can utilize further optimization potential. It must be noted that for practical planning tasks, the oversight and—if necessary—intervention within the entire planning process from manual planners with lots of experience is still advisable, since not all current information can be economically modeled in the automatic planning methods.

In an additional scenario, flexible spot market prices were considered for electricity and provided as input for the optimization module. The feasibility of using the method to synchronize industrial energy demand with fluctuating energy availability was shown. However, in this specific scenario, the additional optimization potential was limited, amounting to ~1% of the energy costs. This is largely due to (i) a limited

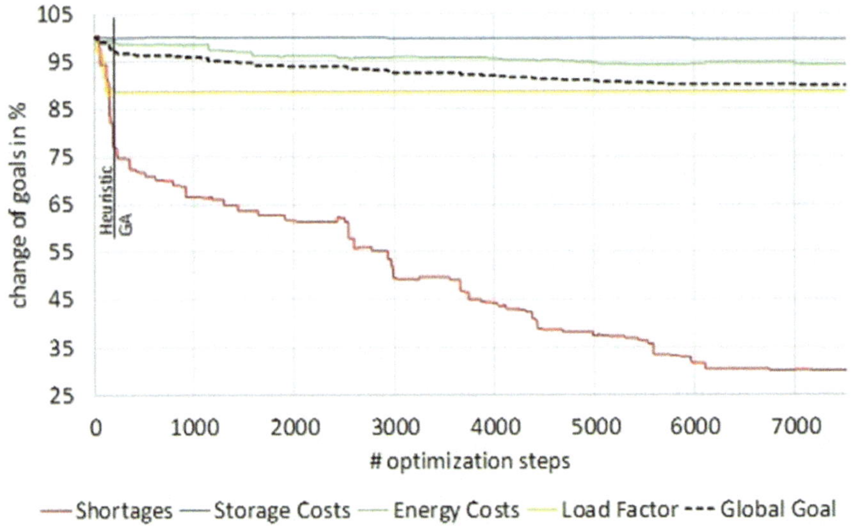

Fig. 2.8 Optimized planning—goal trends (Sobottka et al. 2020), GA: genetic algorithm

lever, as only one of the furnaces is electric, (ii) low energy prices compared to the other sub-goals, and (iii) the circumstance that for the considered time the furnaces were utilized for almost the entire available on-shift time, thus leaving very little room for micro adjustments in the production schedule. Nonetheless, the overall potential for such a digitized automatic energy synchronization is significant, considering the large number of companies that currently do not coordinate their energy use with energy providers on the one hand and the challenges of the global energy transition towards renewables on the other hand (cf. Sauer et al. 2019).

The hybrid simulation provides energy evaluation of the production plans and does not rely on fixed energy consumption profiles, but models the time-dependent interactions between material and energy flows and can include energy exchanges among production equipment as well as with its periphery, e.g., building areas and technical building services (TBS). For example, loading the furnaces must wait for pre-heating to be completed, which in turn depends on the starting temperature and the current outside temperature. With the simulation, the optimization can consider a variety of technological as well as organizational restrictions in real-world application environments. Therefore, it is well suited for practical applications compared to more theoretical approaches from operations research. The integrated simulation and optimization approach, implemented in a rolling horizon planning, could be the core element of an energy-aware digital twin for complex production systems.

2.4.2 Metal-Machining Process Chains

With respect to the ever increasing need for a transition towards an environmentally sustainable economy, manufacturing companies need to incorporate sustainability-related goals in the operation and especially the planning of manufacturing facilities. Acknowledging the manifold interrelationships among factory elements and trade-offs between goal criteria, the integrated evaluation of technical, economic, and environmental objectives remains a challenging task. This case study covers the energy-oriented planning of a highly automated crankshaft production line within automotive component manufacturing. Further information can be found in Labbus (2021), Schmidt (2021), and Labbus et al. (2018).

2.4.2.1 Objectives

Figure 2.9 displays a reference planning process and positions methods and tools to respective planning phases of the crankshaft production line (Schmidt et al. 2017; Labbus et al. 2018). During concept planning, the energy value stream method is applied for line balancing, estimating the production´s energy demand as well as the product´s energy intensity. With respect to the available planning data and the existing uncertainties during this phase, static calculations are sufficient. During detailed planning, energy-oriented modeling and simulation is applied, in order to account for the dynamic interactions inside the process chain. Thereby, a feedback loop from the operation stage (technology analysis) traces back energy, media, and operation data to the previous planning phases. The simulation can either be applied for planning of new process chain configurations or investigating different improvement measures on an existing production line.

As indicated in Table 2.3, machining processes (e.g., turning, grinding) are predominant in the crankshaft process chain. However, an energy-intensive heat treatment process is involved as well.

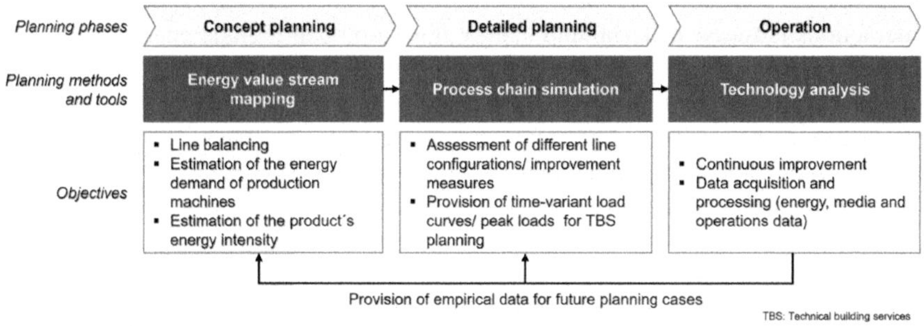

Fig. 2.9 Methods and tools for the energy-oriented planning of process chains, based on Schmidt et al. (2017) and Labbus et al. (2018)

Table 2.3 Process chain of crankshaft production

Nr	Process step
10	Cutting into length, centering
20	Turn broaching
30	Turn milling
40	Boring
50	Induction heating and relaxation
60	Boring
70	External cylindrical grinding
80	External cylindrical grinding
90	Boring, threading
100	Face grinding
110	External cylindrical grinding
120	Boring, internal cylindrical grinding
130	Balancing
140	Boring
150	Finishing
160	Measuring
170	End washing

2.4.2.2 Setup of the Simulation Approach

The underlying logic of the modeling and simulation approach is shown in Fig. 2.10 (Dér et al. 2022). The top level of the approach encompasses a generic process chain, which is composed of a sequence of process steps and corresponding buffers. Each element of the process chain can flexibly be adapted to the current planning case. The machine level reflects a generic behavior of the production machines by differentiating between characteristic states and transitions between them. Here, the processing time and state-based (empirical) power demands are of special importance. As a result, the electrical load curve is determined and value-adding and non-value-adding energy demands are calculated. At process chain level, the single load curves of machines are superposed to the aggregated load curve of the process chain. Other relevant performance indicators are also aggregated on process chain level, e.g., the cycle time and utilization of process chain elements. The modeling and simulation approach is implemented in a three-step solution. The parametrization of the model takes place on a standard spreadsheet document. The simulation model was developed in AnyLogic 8 Professional and exported as a standalone Java application. The results of a simulation experiment are exported to a standard spreadsheet document, where they can be further processed and integrated into other planning tasks. With this procedure, the simulation model is easily accessible for planners, even without modeling and simulation experience.

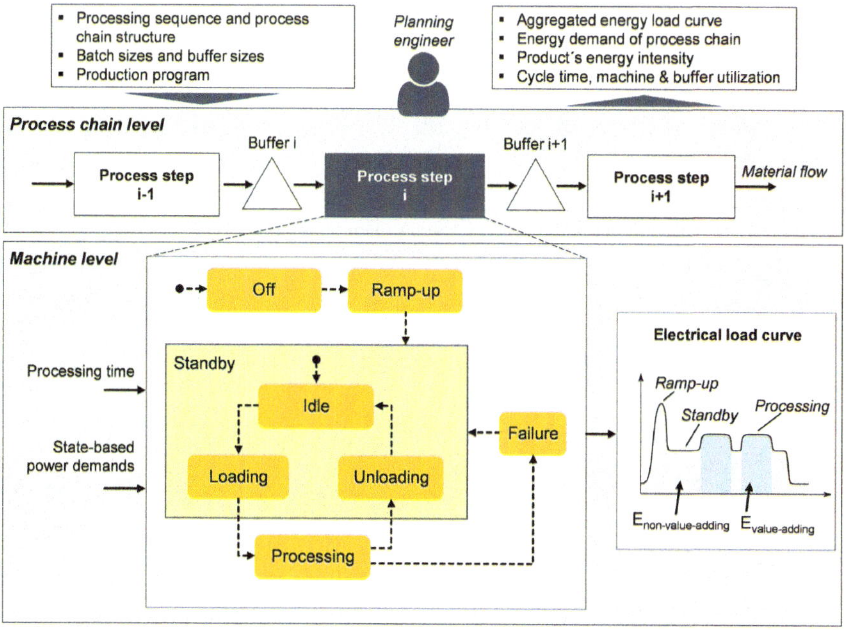

Fig. 2.10 Process chain simulation approach (based on Dér et al. 2022)

2.4.2.3 Results

As the basis for the simulation, the energy value stream methodology is applied first. The goal is to achieve a well-balanced cycle time at all process steps and gain an understanding about the bottlenecks and energy hotspots (Fig. 2.11). At most process steps in this case, a single machine can achieve the line takt of 60 s. However, a longer processing cycle at some process steps (PS), e.g., PS 20, 30, 70, and 80, necessitates the parallelization of machines.

Figure 2.12 shows the calculated energy intensities in descending order. The respecting power demands were retrieved from the technology analysis module, which contains energy data about multiple already existing crankshaft production lines. The top five process steps (approximately 80% of the total product's energy intensity) are external cylindrical grinding (PS 70, 80, and 110), induction heating (PS 50) and face grinding (PS 100).

The simulation experiment in the next planning phase covered three scenarios. All scenarios simulated a period of one week in a three-shift system with working

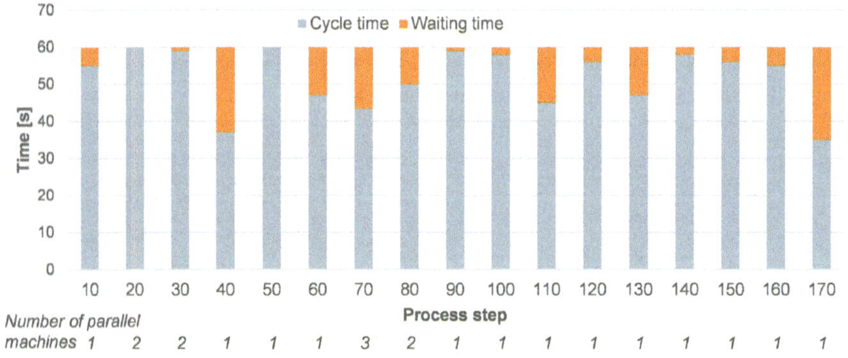

Fig. 2.11 Line balancing within the energy value stream mapping tool

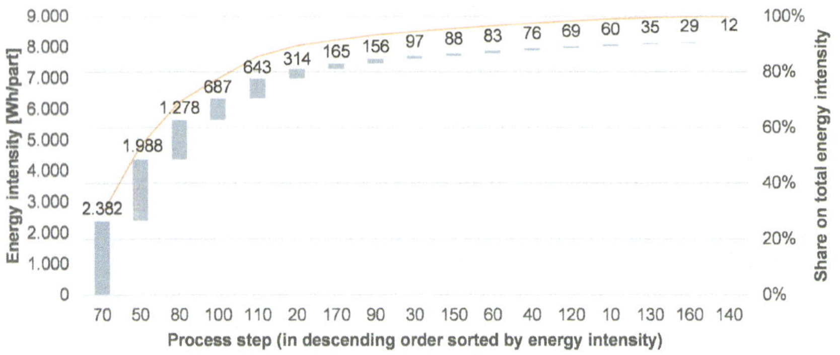

Fig. 2.12 Energy intensity of the crankshaft production

days from Monday to Friday. After defining the scenarios, the simulation model was parameterized via a spreadsheet document that was imported during the simulation run. Figure 2.13 summarizes the simulated scenarios and the corresponding results. Scenario 1 assessed the impact of increasing the processing rate by 5% in the bottleneck process (PS 70). This resulted in a shorter processing time at the machine and a shorter cycle time at the process step. Therefore, the third machine gets dispensable and two machines are sufficient to reach the line takt. The second scenario focused on the reduction of non-value adding energy demands in production-free times. To this end, the standby energy demand was reduced on the machines with the highest standby power demands (PS 70 and 80). Here, best-in-class power demands were retrieved from the technology analysis module. Therefore, this scenario also represents a realistic case. Taking a look at the production volumes, the results demonstrate that all scenarios are equally feasible from a technical perspective. As expected, Scenario 1 stands out with the lowest total energy demand due to the reduced number of machines compared to the remaining scenarios. Scenario 2 suggests that energy efficiency improvements are also achievable, even without changing the process chain structure. The results also indicate that non-value adding energy demands in production-free times add up to a significant share of the total energy demand. Therefore, measures focusing on reducing the base load of machines are effective means of increasing the eco-efficiency of the process chain. However, even more important is the optimal line balancing, which not only saves energy but also improves the technical and economic performance of the process chain.

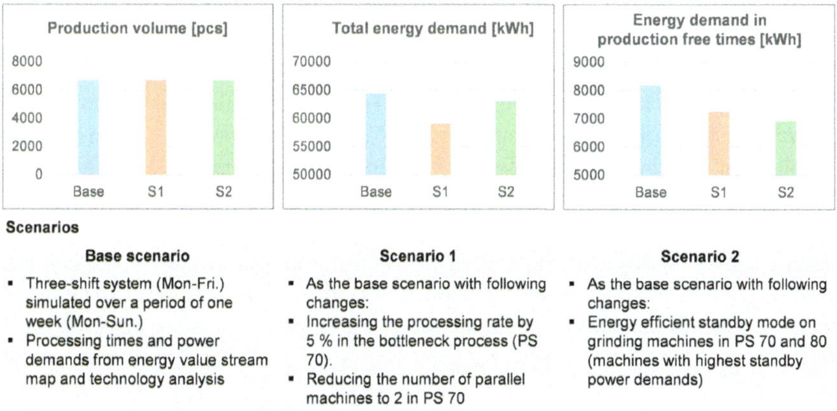

Fig. 2.13 Simulation scenarios and results

2.4.3 Hot Forging

The third case study is provided by a leading producer and supplier of forged power-train components and engineering services in the automotive sector. The company produces in several production sites high-order quantities of individually engineered components by massive forging, averaging at several thousand parts per order. Due to the high processing forces, short cycle times and the heating of products in warm and hot forging up to 1,100 °C, energy costs are a critical part of overall product costs. The manufacturing process consists of raw material separation, blasting, quality checks, several forging stages, as well as heating processes.

2.4.3.1 Process Description and Objectives

A schematic view of the multi-step process from raw material to finished goods can be found in Fig. 2.14. Starting materials are usually bars from raw materials such as steel and aluminum. The bars are first cut into coils to portion the amount of material needed for the next steps. The transport of the bulk material between the machines is carried out manually, the feeding of the machines is conducted automatically. The separated billets are processed by either hot, warm, or cold forging in two steps. In case of warm and hot forging, the billets need to be heated up right before the forging process by an induction unit. Billets are carried through the induction unit to set the right forging temperature within the product. After placing the part in the forging tool, multi-stage forging is carried out. Depending on the product, a second forging operation to set the final geometry, also called calibrating, follows the same process scheme. Before and afterwards, heating and blasting processes create the specific material properties. This particularly improves the compensation of alternating stresses due to continuous fiber orientation compared to solely cut components (Klocke and König 2006). The process is supported by external processing as well as quality checks.

The induction units as major consumers are powered with electricity, but also the forging itself causes additional high energy consumption. Consequently, peak loads as well as overall energy consumption are critical for the final product costs. While the energy efficiency can mainly be manipulated with high invest costs, peak loads refer

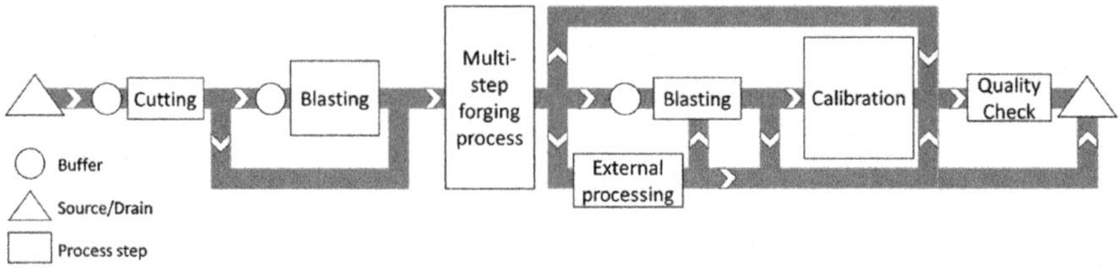

Fig. 2.14 Processing scheme of the real system

heavily to the concept of energy flexibility and subsequently rather organizational measures. Hence, the temporal shift of loads to time frames with either lower overall system load level or – depending on the company's energy purchasing scheme—lower energy costs are appropriate measures. This can be broken down to several planning and control strategies or beneficial energy market interaction. Generally, these targets can be validated within material flow simulation in combination with energy considerations of product preferences and process flow. Hence, transparency in terms of material flow in combination with its energetic profile as well as interdependencies between energy consumption, workload, and processing times need to be investigated to realize flexibility potentials. The implications on systems analysis and simulation modeling will be presented in the following sections.

2.4.3.2 Setup of the Simulation Approach

As stated above, the goal was to model a complex system of forging machines to achieve energetic transparency in the first step. Following the VDI 3633, this refers to the principle "*that a system does not have to be implemented as exactly as possible in a model but as exactly as necessary for the specified investigation aims*" (Verein Deutscher Ingenieure 2014, p. 22). Following this approach, the first step of the modeling consisted of an analysis of energy consumers in relation to the total energy consumption of the factory system. This analysis revealed that the forging processes with the connected induction lines are responsible for about 85% of the electric energy consumption within the system, 15% refers to blasting and separation processes as well as a base load by heating, ventilation, air conditioning, compressed air, and office buildings. Consequently, the simulation itself concentrated on the modeling of the forging machines, while the base load was simulated as a black box with slightly fluctuating consumption. Additional resources such as battery storages or decentralized energy supply were not considered within the model.

Subsequently, the material flow properties of the relevant system elements were analyzed. This included the separation of relevant process steps into their consecutive increments as well as the abstraction of the process to discrete events. Figure 2.15 shows the material flow object to be modeled within the simulation engine.

The forging process can be separated into an infeed, a separation process, the induction conveyor that heats up and transports the parts into the forging process,

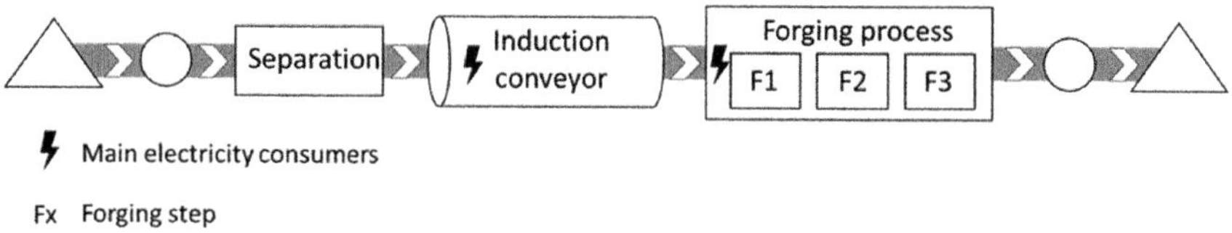

Fig. 2.15 Scheme of a material flow object of the multi-step forging process for the simulation model

as well as the forging process itself, which consists of three single steps. Due to the parallel machining, this can be assumed as one single process step. The finished parts are stored in buffered transport boxes before being passed on to the next step.

As the next step, system load data, organizational data as well as additional technical data were clustered and condensed for the observation period of one month. System load data included data of order input and product data such as process plans and sequences. Organizational data refer to shift models or restrictions within work allocation. As already stated, load peaks are a main restriction that need to be avoided. Technical data comprise information about factory structure and allocation of flow functions to system elements, machine, and corresponding energy data. The acquisition of energy data was focused on the main energy consumers, which are the induction lines as well as the forging stages themselves (Fig. 2.15).

The simulation was implemented with Siemens Plant Simulation 16.1. The material flow is modeled as a job shop, the machines are not arranged in sequence of the working step and there is manual transport between manufacturing steps. The system borders are set with the incoming material from the warehouse and the dispatching area limits the model to subsequent processes. The material flow consists only of forging processes. Due to the highly manual handling and short work plans, the material transport between the processes is considered without a transport time. Figure 2.16 shows the main network of the model including the generic forging process network above the layout.

Forging processes have a recurrent design, which allows for modeling every machine by a generic design and material flow logic. In total, ten forging machines are modeled within the simulation with similar designs.

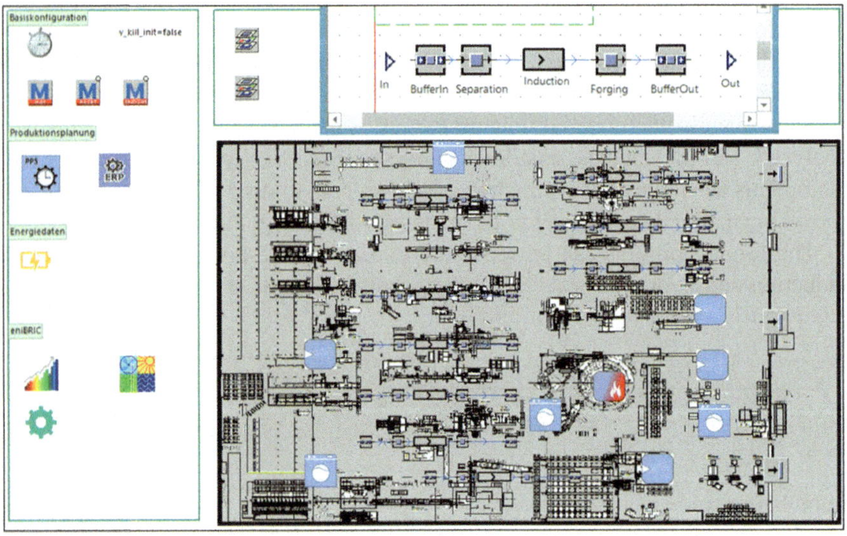

Fig. 2.16 Implemented model in plant simulation (screenshot)

As already stated in Fig. 2.4, the simulation of energy flows within the model is realized by a discrete event simulation and evaluation within one application (paradigm C) for basic investigation. The internal evaluation is realized with the eniBRIC module, a generic library for energy and resource consumption analysis, which is connected to the production resources in the simulation model. It enables the coupling of material flow objects with a module to measure energy consumption (Stoldt et al. 2013), but also to model renewable energy sources and energy storages (Stoldt et al. 2017). By focusing on energetic transparency in material flow simulation, only the former functionality was applied and will be discussed further. The module allows for configuring specific processing states with the corresponding energy consumption for each product and processing station. The eniBRIC module was implemented into the main network to simulate the energy source and enable energy consumption tracking. Several modules were connected to the forging processes to track energy consumption according to their individual processing states. Furthermore, order data, energy data, and technical data were stored centrally to control the order dispatch and product-based configuration of the stations during simulation. Blasting, cleaning, and heating were modeled as black boxes due to the minor influence on total energy consumption.

2.4.3.3 Results

To prepare the simulation for further investigations in terms of energy flexibility measures and energy market interaction, a basic comparison of energetic profiles of the real system and simulation was conducted. A period of one month was chosen due to the provided detailed energy measurement and manufacturing data for this time frame. Both datasets were matched by time stamps to create a load curve for production. Time stamps of manufacturing data were also used to dispatch jobs to the machines during simulation. The energy data were implemented for each job, state, and machine according to the analysis in the last section. Each state change of forging machines in the simulation also triggered a change in energy consumption. Figure 2.17 shows an extracted view on the comparison between simulation and the real system.

The two series of data show a good overlap in time as well as in the level of load. Peaks could be reproduced in satisfactory quality; the processing time and consumed energy of machines matches the real system data. A hurdle to better quality of simulation is the missing data of machine breakdowns in the real system. Hence, peaks can be reproduced in their level, but—depending on the availability of machines—not at a specific time along the entire simulation. This behavior is intensified by variable setup times of forging presses, which lead to delayed job processing. To completely understand the combination of technical and organizational processing in the system, further on-site assessments of the real system need to be conducted. Nevertheless, the effect of job combinations on the machines to peak loads was measurable and reproducible by simulation.

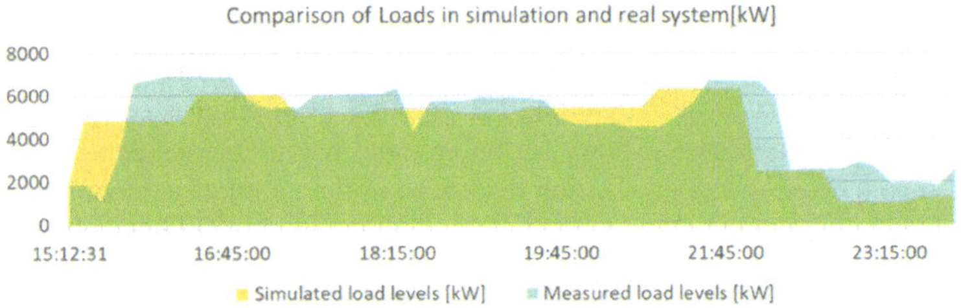

Fig. 2.17 Comparison of loads in the real system and the simulation model

To get further insight into the energetic behavior, the load profiles were aggregated to compare machines with each other. According to Herrmann et al. (2013), a Pareto analysis (Fig. 2.18) as well as an energy portfolio (Fig. 2.19) were depicted to identify hot spots for energy measure application. These clusters structure the impact of the consumers in which.

- Cluster I includes critical machines with high load and working time,
- Cluster II includes machines with low working time but high load levels,
- Cluster III includes machines with high working time and moderate load levels, and
- Cluster IV includes machines with non-critical attributes.

This also helps to determine the criticality of the machines according to their load levels and working times. One hot-forging machine is close to the critical Cluster

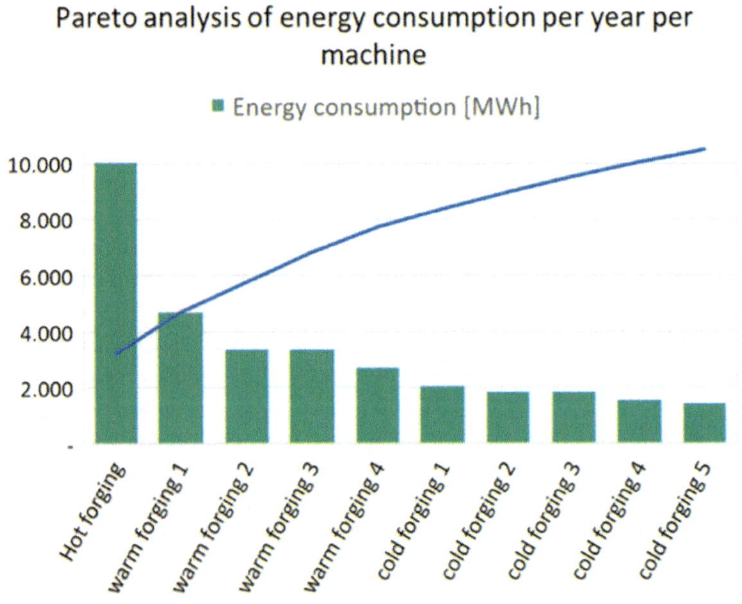

Fig. 2.18 Pareto analysis of energy consumption per year

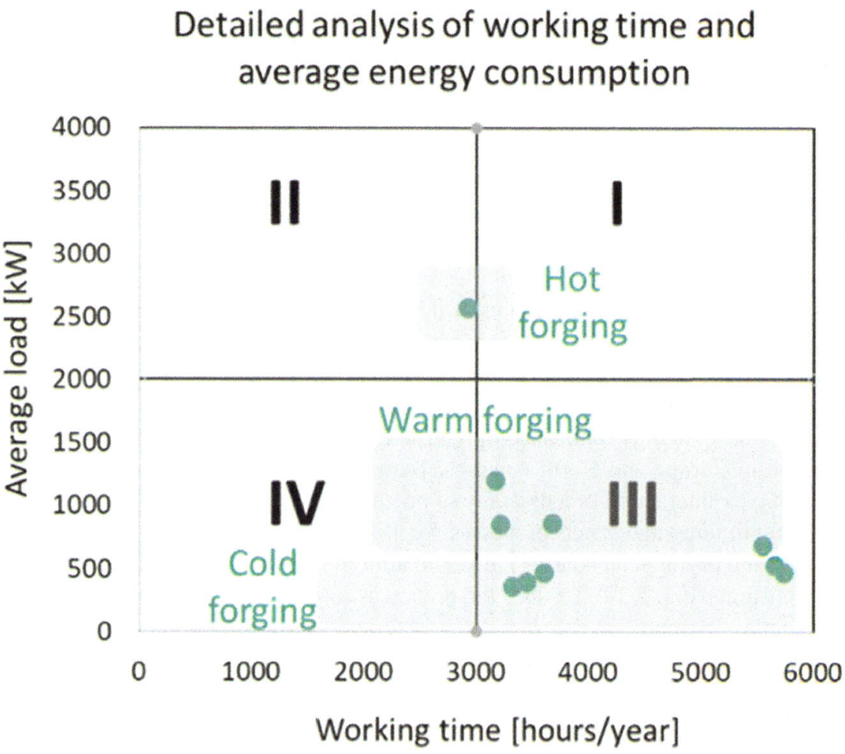

Fig. 2.19 Portfolio analysis of working time and average energy consumption

I and or a driver of peak loads. Subsequently, the hot forging machine needs to be supported by appropriate energy measures. This makes it inevitable to implement further production planning strategies such as a change of the order sequence to avoid combinations of jobs with high energy consumption, shifting the start of orders to delay the parallel processing of energy-intensive jobs or adjusting processing speeds to reduce load levels of the process (Graßl, 2014). Incremental shutdowns of forging processes might also be an option to avoid load peaks in the real system.

Furthermore, static energy purchasing schemes may be replaced by dynamic ones through the consideration of energy costs during factory operations. Energy flexibility offers at the day-ahead and intraday market might be another option to reduce overall energy costs. By planning the above-mentioned load reductions, an additional flexibility offer may be placed at the energy markets to generate further value from peak load reduction. This is underpinned by the high production flexibility within the system due to many manual transport, setup, and maintenance processes in combination with short production plans and redundant machines. A further step in the model development will be the implementation of algorithms for peak shaving by increasing the processing times per part. This increases the overall cycle time, but also reduces the energy consumption per part and might be a step towards more flexibility in manufacturing control. Also, algorithms for flexible scheduling of jobs in order to decrease the parallel processing of energy-intensive jobs will be tested and validated in the simulation model. This also requires to set up a database of product-specific energy consumptions per machine state and machine. Furthermore, a comparison of load levels during daytime and actual day-ahead and intraday energy prices can be implemented to identify time slots with cost saving potential.

2.4.4 Battery Cell Production

The last case study addresses an industrial-scale research factory where all manufacturing steps are available in order to produce automotive-oriented pouch battery cells. Batteries play a crucial role, e.g., for the energy transition of mobility. The demand for battery cells is estimated to grow significantly in the next decades. While battery manufacturing mainly took place in Asia so far, more and more battery factories are planned in Europe and North America. Battery manufacturing is characterized by the complex interaction of a diversity of processes (Kwade et al. 2018). This ranges from continuous and batch processes for the electrode production (e.g., mixing, coating and drying, calendaring) to single unit processes for the later cell assembly and finishing (Fig. 2.20). Besides the processes as such, technical building services (TBS) play an important role in battery factories. The most important aspect in this context is the necessary dry room environment for the cell assembly, which is typically also a main driver for the energy demand (Thomitzek et al. 2019). Details of the case study are reported by Schönemann (2017) and Schönemann et al. (2019).

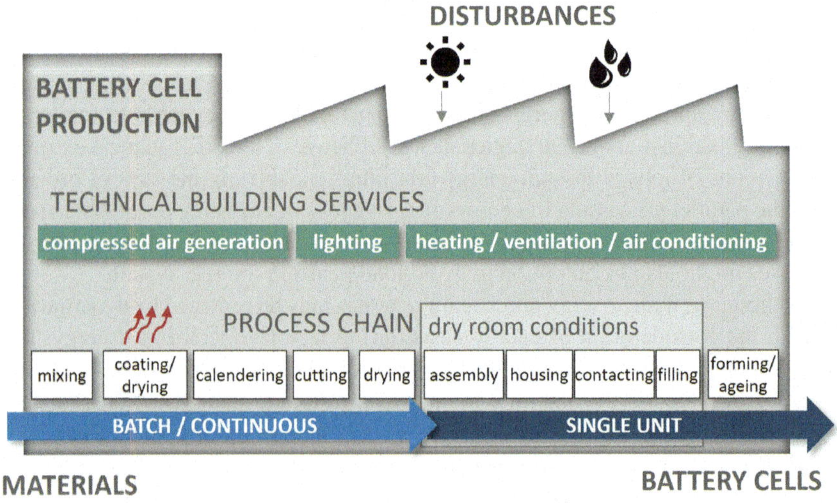

Fig. 2.20 Battery cell factory—overview of elements (Thiede et al. 2019)

2.4.4.1 Objectives

Energy demand of battery manufacturing can actually be seen as a crucial success factor. It mainly determines the carbon emissions (Ellingsen et al. 2014) and is also relevant from the cost perspective; clearly behind material costs, but in the same order of magnitude as labor or depreciation of investment. Therefore, the objective of the simulation study is to support the planning of battery factories explicitly including energy aspects.

2.4.4.2 Setup of the Simulation Approach

Given the heterogeneous setup and the complex interactions in a battery factory, a hybrid agent-based simulation model with discrete event and continuous simulation techniques and a simulation architecture with hierarchical coupling of different models (related to paradigm B in Fig. 2.5) was developed (Schönemann 2017). Figure 2.21 gives a simplified overview of the main components. The core model pursues an agent-based approach that was realized in AnyLogic. It depicts the manufacturing process chain with all machines, the material flow (product units moving through the system), and also workers.

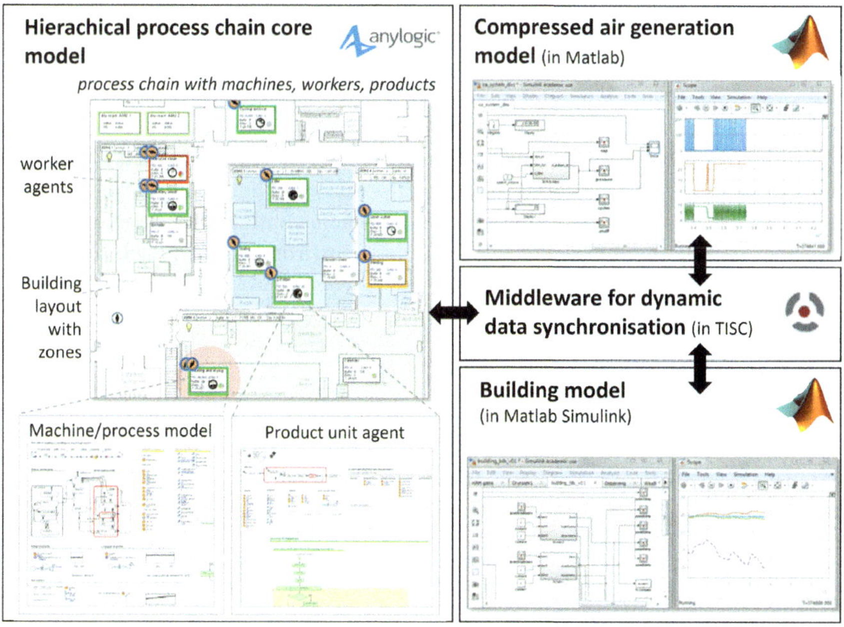

Fig. 2.21 Multiscale simulation approach with coupling of different simulation tools (adapted from Schönemann 2017; Schönemann et al. 2019)

All those entities incorporate specific models to depict their individual behavior. Therewith, also specific machine equipment and different types of processes can be considered in the necessary level of detail. Additionally, different production zones (e.g., dry room, other air conditioned environment) are distinguished. For the technical building services, separate models in Matlab Simulink, e.g., for compressed air generation and also the building with its different zones, were realized. This enables a more-detailed representation of the physics and thermodynamics, e.g., when it comes to temperatures in different zones and also energy demand for the related HVAC system. All those models are coupled and orchestrated through a middleware (TISC). With this overall architecture, complex interdependencies among all entities can be well considered. As one example, machine operation or worker presence lead to heat loads in the building models and, thus, to higher energy demand for keeping the required ambient conditions. External, e.g., seasonal, influences can be considered, too. Another example is the realistic consideration of compressor operation and related energy demand based on machine activities.

2.4.4.3 Results

Figures 2.22 and 2.23 show exemplary energy-related results of the presented simulation architecture (Schönemann et al. 2019). Based on the dynamic simulation of activities within the battery factory, detailed energy load profiles of all different elements can be derived (Fig. 2.22). This gives deep insight into energy-related interdependencies and enables the identification of improvement measures as well as the analysis of their potential impact.

Figure 2.23 provides the exemplary breakdown into the main energy drivers of the different factory elements. Results underline the very strong relevance of technical building services and the coating and drying processes. The simulation environment allows for systematically deriving alternative system configurations and operational strategies in order to improve the energy-related environmental impacts and costs.

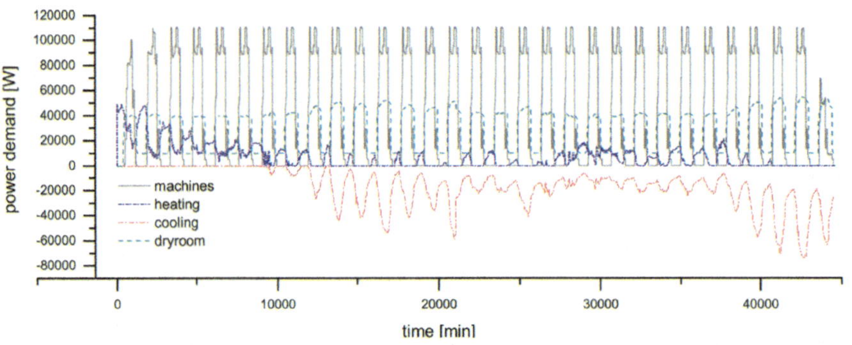

Fig. 2.22 Exemplary load profiles as result of energy simulation of battery manufacturing (adapted from Schönemann et al. 2019)

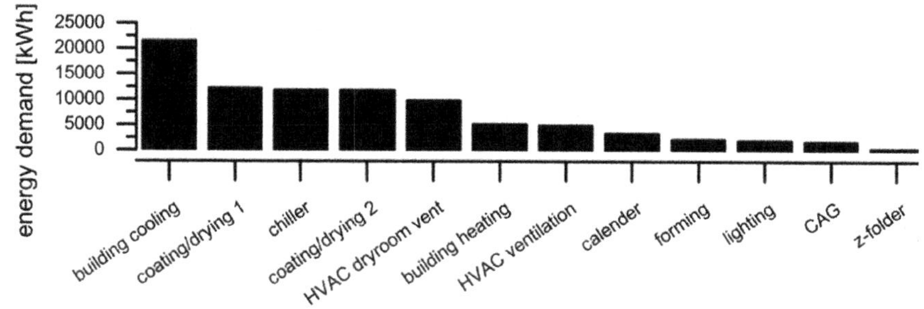

Fig. 2.23 Exemplary breakdown of energy demand to processes as result of energy simulation of battery manufacturing (adapted from Schönemann et al. 2019)

2.5 Conclusions and Outlook

This chapter has underlined the feasibility and the potentials of energy-oriented simulation approaches for manufacturing systems. The necessary underlying methods (e.g., for modeling the energy demand of machines) are well established and a diversity of implementations with different focusses and architectures are available in research as well as in commercial solutions. As demonstrated in the case studies, both energy efficiency and flexibility aspects can be addressed. Therefore, energy-oriented manufacturing system simulation can make a significant contribution towards the energy transition of individual companies and the manufacturing industry as a whole. There are, of course, still a couple of challenges to be overcome for broader application as well as further areas for future research can be identified. Without doubt, the continuous acquisition and target-oriented utilization of energy data can be further improved. While in general the technical and methodological means are existing, many studies are still based on temporally limited measurements, or consistent data in the right spatial and temporal resolution are not sufficiently available. The ongoing digital transformation of manufacturing industry will certainly further facilitate to overcome this barrier. More-advanced and economically feasible sensors and further IT hardware as well as more-standardized IT architectures and related interfaces (e.g., OPC UA) are available. This facilitates the acquisition, storage, and processing of data as well as the combination of different data sources, and eventually lead to more beneficial use cases. In this context, also the clearer definition of necessary resolutions of energy data inputs for different applications is interesting. The goal is a balance of the necessary accuracy for achieving potential improvements with the related efforts for data acquisition and processing. The continuous availability of data in combination with appropriate models also enables the development of digital twins as up-to-date digital representatives of the physical system. Those approaches facilitate advanced use cases with a broader scope beyond planning towards continuously supporting the operation of manufacturing systems, e.g., through energy-oriented control. In this context, also the combination with data-based methods (e.g., based on machine learning) is promising, which could be interesting for automated model generation as well as advanced assessment of simulation results.

Regarding the lesson learnt, one underlying challenge in manufacturing is certainly that energy-related questions were often perceived as less urgent and beneficial compared to other areas, which lead to a lower prioritization on these issues. Given the environmental relevance and the increasing economic dimension of energy demand, this needs to change in the future—energy-related aspects should be a regular part of manufacturing system improvements, like nowadays time, quality, and output-oriented indicators. Besides improving the feasibility of specific energy-related use cases as pointed out above (increasing the benefits, e.g. through prioritization) and focusing on decreasing efforts (e.g., through easier applicability), tapping the synergies with other applications is a promising approach to further facilitate those developments. Considering energy flows in manufacturing systems should not be a side issue that is just considered by specific experts. It needs to be integrated into normal routines in both planning and operation, while making use of latest advancement in the ICT domain (cf. Labbus 2021). Modeling and simulation can be an important enabler towards deeper understanding of interdependencies, forecasting of future scenarios and derivation of meaningful improvement pathways.

References

Dehning P, Blume S, Dér A, Flick D, Herrmann C, Thiede S (2019) Load profile analysis for reducing energy demands of production systems in non-production times. Appl Energy 237:117–130

Dér A, Kaluza A, Reimer L, Herrmann C, Thiede S (2022) Integration of energy oriented manufacturing simulation into the life cycle evaluation of lightweight body parts. Int J Precis Eng Manuf-Green Technol 9:899–918

Ellingsen LAW, Majeau-Bettez G, Singh B, Srivastava AK, Valøen LO, Strømman AH (2014) Life cycle assessment of a lithium-ion battery vehicle pack. J Ind Ecol 18(1):113–124

Gebler M, Cerdas JF, Thiede S, Herrmann C (2020) Life cycle assessment of an automotive factory: identifying challenges for the decarbonization of automotive production—A case study. J Clean Prod 270:122330

Graßl MA (2014) Bewertung der Energieflexibilität in der Produktion. Ph.D. thesis, TU München, Werkzeugmaschinen und Betriebswissenschaften, München

Herrmann C, Thiede S, Kara S, Hesselbach J (2011) Energy oriented simulation of manufacturing systems—concept and application. CIRP Ann 60(1):45–48

Herrmann C, Posselt G, Thiede S (2013) Energie- und hilfsstoffoptimierte Produktion. Springer, Berlin Heidelberg

IPCC (2014) Climate Change 2014: mitigation of climate change. Contribution of working group iii to the fifth assessment report of the intergovernmental panel on climate change. Cambridge University Press, Cambridge

Klocke F, König W (2006) Fertigungsverfahren: Umformen. Springer, Berlin Heidelberg

Kwade A, Haselrieder W, Leithoff R, Modlinger A, Dietrich F, Droeder K (2018) Current status and challenges for automotive battery production technologies. Nat Energy 3(4):290–300

Labbus I (2021) Cyber-physische Produktionssysteme für die energieeffiziente Komponentenproduktion. Springer Fachmedien, Wiesbaden

Labbus I, Schmidt C, Dér A, Herrmann C, Thiede S (2018) Automated production data integration for energy-oriented process chain design. Procedia CIRP 72:551–556

Sauer A, Abele E, Buhl HU (2019) Energieflexibilität in der deutschen Industrie: Ergebnisse aus dem Kopernikus-Projekt – Synchronisierte und energieadaptive Produktionstechnik zur flexiblen Ausrichtung von Industrieprozessen auf eine fluktuierende Energieversorgung (SynErgie). Fraunhofer Verlag, Stuttgart

Schmidt C (2021) Planning of eco-efficient process chains for automotive component manufacturing. Springer, Berlin Heidelberg

Schmidt C, Labbus I, Herrmann C, Thiede S (2017) Framework of a modular tool box for the design of process chains in automotive component manufacturing. Procedia CIRP 63:739–744

Schönemann M (2017) Multiscale simulation approach for battery production systems. Springer International Publishing, Cham

Schönemann M, Bockholt H, Thiede S, Kwade A, Herrmann C (2019) Multiscale simulation approach for production systems. Int J Adv Manuf Technol 102(5):1373–1390

Sihn W, Sobottka T, Heinzl B, Kamhuber F (2018) Interdisciplinary multi-criteria optimization using hybrid simulation to pursue energy efficiency through production planning. CIRP Ann 67(1):447–450

Sobottka T, Kamhuber F, Rössler M, Sihn W (2018) Hybrid simulation-based optimization of discrete parts manufacturing to increase energy efficiency and productivity. Procedia Manuf 21:413–420

Sobottka T, Kamhuber F, Heinzl B (2020) Simulation-based multi-criteria optimization of parallel heat treatment furnaces at a casting manufacturer. J Manuf Mater Process 4(3):94. https://doi.org/10.3390/jmmp4030094

Stoldt J, Schlegel A, Franz E, Langer T, Putz M (2013) Generic energy-enhancement module for consumption analysis of manufacturing processes in discrete event simulation. In: Nee A, Song B, Ong SK (eds) Re-engineering manufacturing for sustainability. Springer, Singapore, pp 165–170

Stoldt J, Prell B, Rabe M, Wenzel S, Thiede S (2021) A criteria-based database for research and applications of energy-oriented simulation in production and logistics. In: Schuderer P, Franke J (eds) Simulation in Produktion und Logistik Cuvillier, Göttingen Germany, pp 93–102

Stoldt J, Prell B, Schlegel A, Putz M (2017) Modellierung von volatilen erneuerbaren Energieerzeugern und Energiespeichern in Siemens Plant Simulation. In Wenzel S, Peter T (eds) Simulation in Produktion und Logistik 2017. Kassel University Press, Kassel

Thiede S (2012) Energy efficiency in manufacturing systems. Springer, Berlin Heidelberg

Thiede S (2021) Digital technologies, methods and tools towards sustainable manufacturing: does Industry 4.0 support to reach environmental targets? Procedia CIRP 98:1–6

Thiede S, Seow Y, Andersson J, Johansson B (2013) Environmental aspects in manufacturing system modelling and simulation—state of the art and research perspectives. CIRP J Manuf Sci Technol 6(1):78–87

Thiede S, Turetskyy A, Kwade A, Kara S, Herrmann C (2019) Data mining in battery production chains towards multi-criterial quality prediction. CIRP Ann 68(1):463–466

Thomitzek M, von Drachenfels N, Cerdas F, Herrmann C, Thiede S (2019) Simulation-based assessment of the energy demand in battery cell manufacturing. Procedia CIRP 80:126–131

Verein Deutscher Ingenieure (2014) VDI Guideline 3633 Simulation of systems in materials handling, logistics and production—Part 1. Fundamentals. Beuth, Berlin

Walther J, Weigold M (2021) A systematic review on predicting and forecasting the electrical energy consumption in the manufacturing industry. Energies 14(4):968. https://doi.org/10.3390/en14040968

Chapter 3
Automotive

Tim Peter, Kristina Sokoll, Wolfgang Schlüter, and Johannes Dettelbacher

Abstract The automotive industry is an important branch in many industrialized countries in Europe, the USA, Japan, and China. Recent political developments provide challenges to achieve carbon–neutral production and to switch from combustion engines to electric engines or other alternative fuels. This is also reflected in the development of simulation in production and logistics, where energy-related questions became more present in recent years. In this chapter, data from the ASIM working group about related literature are analyzed. Topics and architectures already being examined in energy simulation in the automotive industry are reviewed. The developments are illustrated by two application cases. The first use case shows the importance of simulating the power consumption and charging strategies of automated guided vehicles on the performance of the material flow system. The second example shows the possibilities of integrating complex thermodynamic processes into the material flow simulation by combining continuous and discrete models. The chapter ends with a short conclusion on the state of the art regarding energy simulation in the automotive sector.

T. Peter (✉)
Gstädtstraße 30, 82439 Großweil, Germany
e-mail: mail@tim-peter.de

K. Sokoll
AUDI AG, 85045 Ingolstadt, Germany
e-mail: kristina.sokoll@audi.de

W. Schlüter
Stolper Straße 23, 90451 Nürnberg, Germany
e-mail: wolfgang.schlueter@hs-ansbach.de

J. Dettelbacher
Zeitzenhofer Straße 10, 96182 Reckendorf, Germany
e-mail: johannes.dettelbacher@hs-ansbach.de

S. Wenzel et al. (eds.), *Energy-Related Material Flow Simulation in Production and Logistics*, https://doi.org/10.1007/978-3-031-34218-9_3

3.1 Introduction

The production of cars, trucks, busses, and other motor vehicles can be summarized under the term "automotive". However, the production of a car requires several steps and various up- and downstream processes in the value chain, displayed in Fig. 3.1. The actual assembly of a car in the Original Equipment Manufacturer (OEM) demands upstream processes for the chassis of the vehicle. Some of them are especially energy consuming. The following examples show the link between production steps and energy key performance indicators:

Flat steel is shaped and cut into chassis parts in the press shop. Different chassis parts are assembled into the car body in the body shop. New developments lead to the casting of large body parts (Visnic 2020) that replace the classic press shop parts. Furthermore, important structural parts that require exceptional strength in the case of an accident are hardened in hot forming processes, where the parts are heated before they go to the press shop, leading to additional strength and rigidity. Hot forming as well as casting of large body parts are very energy-intensive processes (Geckler et al. 2020, p. 305).

The completed car body is transferred to the paint shop, where set-up times and strategies are important topics. They have to be considered in the planning of the sequence, while also respecting drying times, which depend on temperature and humidity.

The manufacturing of the chassis is a highly automated process with excessive usage of robots for welding and assembly. Simultaneous acceleration of robots leads to undesirable short peak loads in energy consumption.

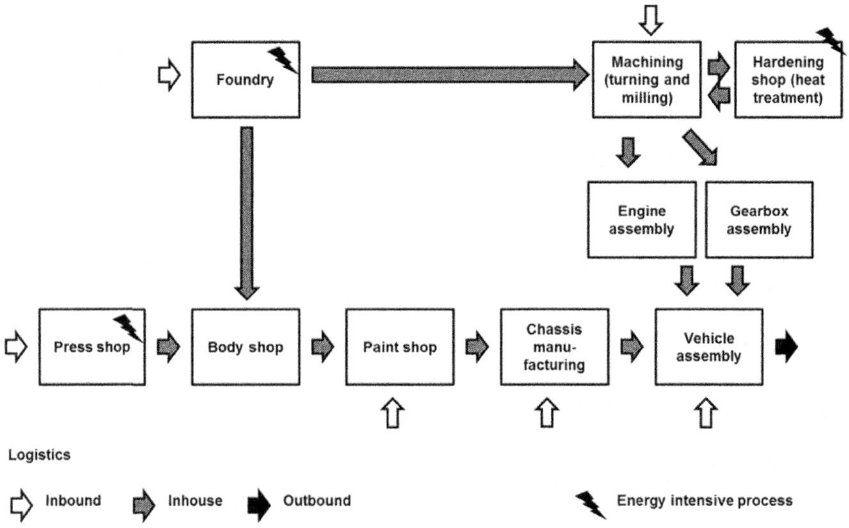

Fig. 3.1 The automotive production process chain

After the chassis is manufactured, it is finally moved to the vehicle assembly line where components such as engine (internal combustion engine or electric) and gearbox are united. Gearbox and engine are often manufactured in another factory, either in component factories of the same OEM or bought from external suppliers. Before they are assembled, the engine and gearbox components require shafts and wheels that are machined, often by milling and turning. The raw parts are cast in a foundry and after initial machining hardened by heat in a hardening shop. These processes again require significant amounts of energy.

In conclusion, the automotive production processes are diverse and cover a wide range of energy-intensive and highly automated process steps. The most energy-intensive processes were identified by Geckler et al. (2020, pp. 304–305) as casting, heat treatment, hardening, and providing process heat.

This section has provided an overview about the automotive production process and examples of energy-intensive processes in the value chain that are worth simulating. In Sect. 3.2, the special features of the automotive industry when simulating energy are presented, based on the results of a detailed literature analysis by the ASIM working group (see Sect. 1.1).

3.2 Simulation of Energy Aspects in the Automotive Production

While the simulation of material flows is common and widespread over most of the mentioned processes (cf. Mayer et al. 2020), energy-related-aspects are relatively new and still an academic topic. Though, with the increase in energy prices and the growing awareness for climate change and the need to decarbonize production, especially the energy-intensive processes in the value chain offer potential for including energy-related aspects into the material flow simulation.

In order to decarbonize production, an increased use of renewable energies is required. Wind and solar energy are heavily dependent on the weather and the time of day and year and are not continuously available. This influences the management of energy-intensive processes and offers an area of application for energy simulation (see Sect. 7.5).

Geckler et al. (2020) separate these potentials into three categories: balancing of energy on a factory or project level, simulation of a factory floor on systems level, and detailed simulation of processes and (single) machines.

The analysis of the ASIM working group shows that most of the case studies for energy-related material flow simulation come from the automotive sector (Wenzel et al. 2017). One reason for this is the high degree of automation in the automotive industry, which makes the processes comparatively easy to model. On the other hand, improvements are only possible with great effort. Simulation is a helpful tool to leverage the remaining potential.

The case studies in the literature contain so many different processes that the working group divided them into OEM car manufacturing as well as part and component manufacturing. As explained in Sect. 3.1, part and component manufacturing are sometimes conducted in-house, sometimes in specialized component factories, and sometimes outsourced to a supplier. In the following sections, both are considered equally and treated together.

The goal is to provide the reader with information regarding important questions and applications of energy-related simulation in the automotive industry. Sources for further reading help to get inspiration and ideas for modeling and simulation. Furthermore, subjects are identified that are still not very well researched and lack applications.

3.3 Findings of the ASIM Working Group

The ASIM working group (see Sect. 1.1) has gathered relevant scientific research and case studies for energy simulation in material flow systems and classified them into a database of eleven different categories which are discussed and classified into a morphological box in Chap. 1 of this book. The following analysis uses the update of the database from April 2021.

One of the categories in which the work has been classified is the "industry" in which the consideration of energy has taken place in the simulation. In the following, the publications related to the automotive industry are analyzed in detail. In total, 51 sources were identified for the automotive sector (24% out of a total of 213), which is by far the largest group among all industries. The aim is to show the focus of simulation studies for automotive use cases and discover fields that lack application and research examples.

3.3.1 Scope and Objectives

In the analysis of the ASIM working group, the publications of authors and author teams have been categorized into eleven groups regarding the scopes and objectives of the work. Figure 3.2 shows the distribution of the publications for automotive industry regarding the scope of the scientific work. Some authors pursue multiple objectives, so that the total amount of scopes in the figure exceeds 51.

While there is a wide variety of scopes and objectives pursued, there is a clear priority for.

- Choosing and sizing of energy-related infrastructure (e.g., Tur et al. 2019; Beier et al. 2016; Kuhlmann and Sauer 2019)
- Local optimization of the energy consumption of single machines and systems (e.g., Sinnemann et al. 2020; Alvandi et al. 2015; Wilson et al. 2016)

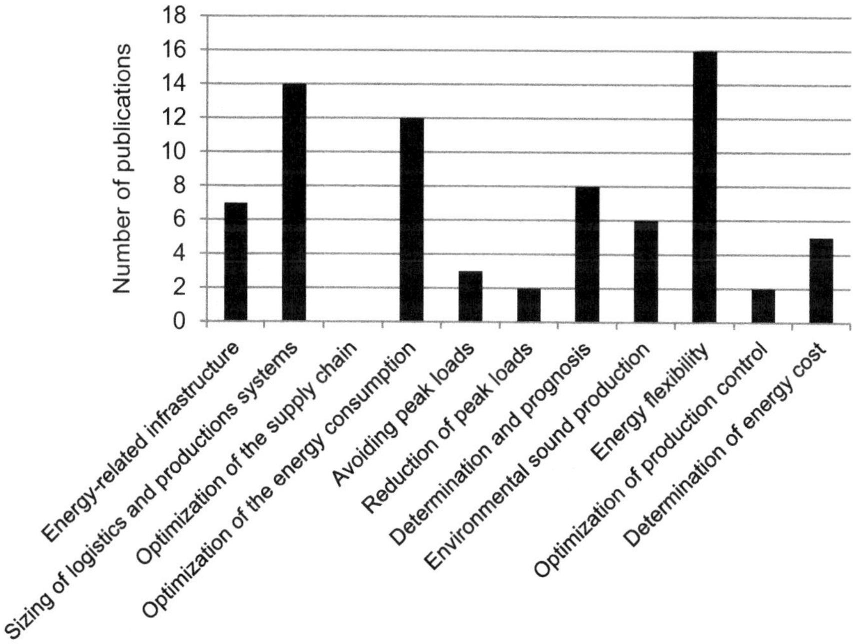

Fig. 3.2 Distribution of automotive publications according to objectives

- Energy flexibility and energy-sensitive production planning and control (e.g., Dunkelberg et al. 2020)

which together make up for 56% of all examined publications in the automotive industry. Examples for relevant publications were chosen with priority to recent works and, if possible, in English language.

3.3.2 Level of Detail of Energy Simulation in the Automotive Sector

Regarding the level of detail, the ASIM working group has identified five types of simulation studies. For the papers in the automotive field, the distribution of these types is shown in Fig. 3.3.

- Most publications deal with simulation on a production line or a production process level (e.g., Stange and Bös 2019).
- Six works concentrate on single components or energy consumers (e.g., Sinnemann et al. 2020; Wenzel et al. 2015; Brüggemann et al. 2014).

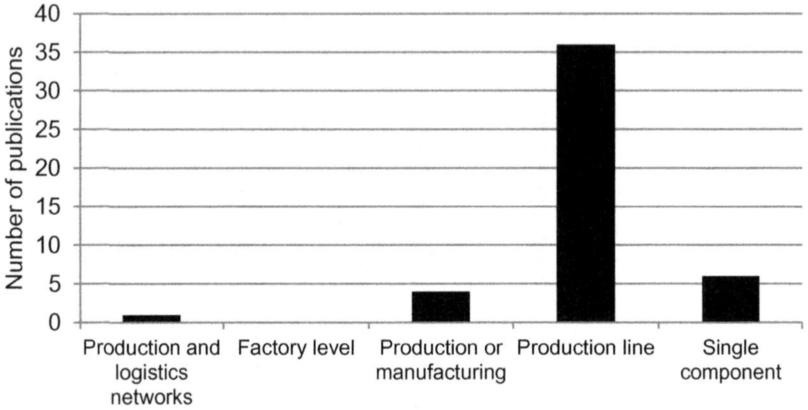

Fig. 3.3 Concentration of papers for the different levels of detail

- Another four works concentrate on simulation on production or manufacturing area level (e.g., Thomitzek et al. 2019; Omar et al. 2016; Ichimura and Takakuwa 2013).
- The only publication describing a use case for production and logistic networks is Lee et al. (2012).
- For the factory level, no publications were identified. Unlike production and manufacturing, the factory level looks at the whole plant, not just a manufacturing or assembly area.

3.3.3 Architecture Used in Energy Simulation in the Automotive Sector

Poeting et al. (2019) displayed four different approaches on how energy can be included in a simulation model. The ASIM workgroup defined further categories for the consideration of energy in the simulation studies (see Sect. 1.2.2).

In the automotive field, by far the most common method is the integration of the energy consideration into the DES simulation tool (e.g.,Sivapragasam 2016; Thiemicke 2016) as shown in Fig. 3.4. It is not only the easiest method to implement, but DES simulation tools are very common in the automotive sector (e.g., Tecnomatix Plant Simulation) and there is significant standardization available (Automotive Library, cf. Sokoll et al. 2021). Both, Plant Simulation and the Automotive Library, contain elements and methods for the simulation of energy consumption of production equipment and transport vehicles.

- Three papers used a discrete event simulation with evaluation after the end of the simulation run (Chu et al. 2016; Wilson et al. 2016; Neugebauer et al. 2012).

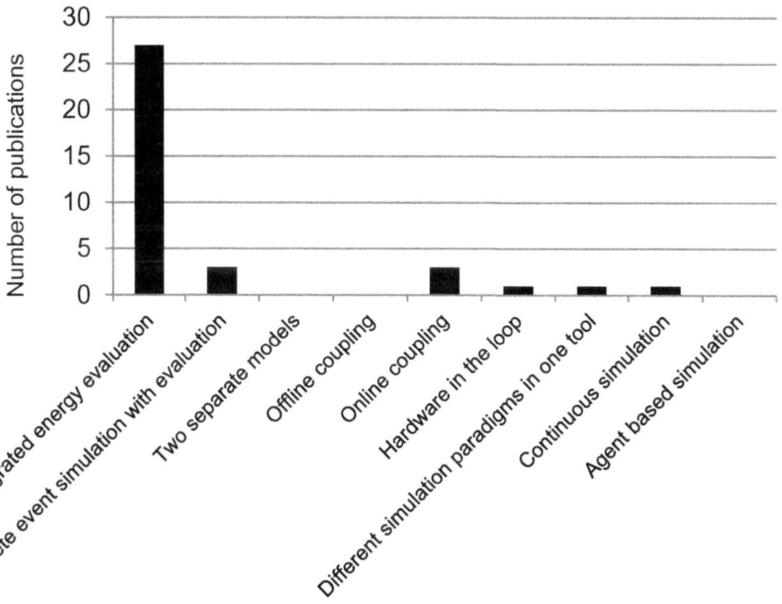

Fig. 3.4 Works regarding the architecture of the simulation models

- Another three ones coupled a discrete event model online with a continuous model (Geckler et al. 2020; Omar et al. 2016; Wenzel et al. 2015).
- One publication each was identified for hardware in the loop (Sinnemann et al. 2020), continuous simulation (Brüggemann et al. 2014), and combining continuous and discrete models in one tool (Thomitzek et al. 2019).

In conclusion, most of the simulation studies in the automotive industry deal with production topics, whereas logistics and transportation are neglected. This seems logical as according to Geckler et al. (2020), 93% of the CO_2 emissions are caused by different production processes and only 7% by "supporting processes" that also include logistics.

3.4 Applications

This will be illustrated by two use cases from fields that are not yet widely covered by simulation. In Sect. 3.4.1, a use case from intralogistics is described. The following use case (Sect. 3.4.2) explains the combined use of discrete event and continuous simulation in one tool, which is an architecture where not many papers were found. Being relatively complex with respect to the modeling task, it is most suited for processes that are described by differential equations, e.g., thermal processes like casting or heat treatment. As these processes are the most energy-consuming, it is worth to analyze them in detail.

3.4.1 Automated Guided Vehicles in Automotive Intralogistics

Automated guided vehicles, hereinafter referred to as AGVs, are increasingly becoming a standard technology in automotive intralogistics. Vehicles as well as necessary material loading and charging infrastructures are becoming more and more performant, reliable, and financially rewarding. Many different configurations for both give the opportunity to put together a specific compatible transport system for each production process.

Therefore, the configuration of logistics systems, varying all those possible characteristics of vehicles, infrastructure, and steering is in the focus of simulation studies to determine an optimal solution. Especially energy-related parameters of vehicle batteries as well as possible energy management strategies (e.g., using stationary or inductive charging) can have a significant influence on the results.

3.4.1.1 Scope of Material Flow Simulation in the Field of Logistics

The overall objective of simulation in the field of transport logistics is to obtain a reliable supply of a production system with a minimum of required resources, such as storage area, transport units, or vehicles. Due to the broad range of attributes describing the logistic process and being variable in simulation experiments, it is possible to reach a predefined target state by many different configurations. The optimization is usually an iterative and non-automated procedure, involving the experience of both simulation and logistics engineers.

The energy of interest in intralogistics is usually the one of battery-electric transport vehicles as a limited resource of the logistic system. Typically, the simulation model is used to find an optimum for the necessary resources, which are not only the required number of AGVs. Related process components, such as material loading and charging stations, have a significant influence on the simulation results. Their number as well as energy-related characteristics and control (e.g., charging amount and logic) determine the main outcome of the simulation study.

The application described here is a simplified use case derived from an early phase of a planning project, describing a supply for an assembly based on automated guided transport vehicles, in order to explain the effects of energy-related attributes on an intralogistics system. Different charging configurations and resources are compared in the simulation experiments. A schematic representation of the examined and typically structured body shop is shown in Fig. 3.5.

For the simulation study, only the storage area for produced components within the body shop, such as doors and lids, and the mounting of those components on the car body at an assembly is of interest. The previous value chain from start of

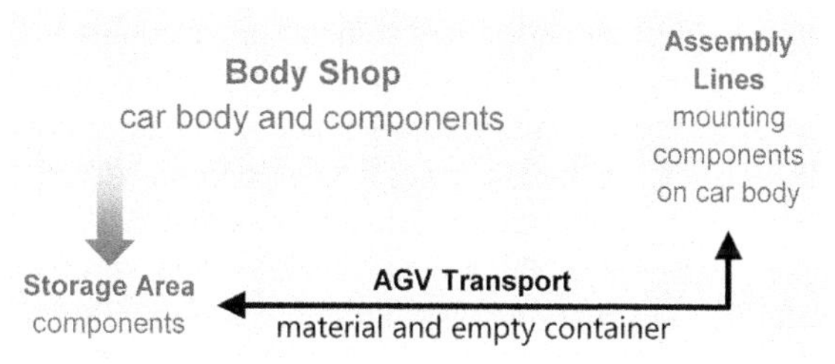

Fig. 3.5 Schematic representation of production and AGV transport of components

production until start of assembly can be modeled in an abstract manner. At the end of the assembly lines, the finished body-in-white is leaving the body shop towards the paint shop.

3.4.1.2 Modeling of Battery-Electric Vehicles

It is observable in recent simulation projects that the higher the degree of automation, the higher is the level of detail of the modeled vehicles. The modeling of conventional vehicles, e.g., manually guided ones such as tuggers or forklifts, include a human factor. Therefore, higher uncertainty and the possible error of the simulation model are accepted and taken into account. In case of automated transports, both vehicles and steering can be described and computed more accurately without additional range of fluctuation caused by a human factor, especially if the transport system is operating within a closed area.

The growing level of detail does not only concern energy, but also other attributes. For manually operated vehicles, such as conventional forklifts or tuggers, the modeled characteristics often include:

- Average driving speed (if necessary, differentiation by normal and slow, e.g., for passing crossroads)
- Loading capacity for transport units and limits (if necessary, both weight and dimension)
- Time needed for handling of transport units (if necessary, differentiation by type of handling, e.g., load or unload)

The technical availability of non-automated vehicles is usually high and may, thus, be neglected. But, for some use cases or kinds of vehicles it might be reasonable to schedule at least fixed maintenance intervals. In addition, it is common practice to simplify energy parameters and controls to a necessary minimum: The battery is assumed to be always sufficiently charged and the energy consumption of the vehicles' actions is not considered. As those kinds of vehicles often have changeable battery systems, lasting for the length of a production shift of eight hours or more, the only relevant influence on the transport disposition is the frequency and time for replacing the currently installed battery with a fully charged one. This is often scheduled at the beginning of each production shift for a maximum number of vehicles at once. For that period less vehicles are available for transportation. Therefore, it is essential to take this kind of organizational unavailability into account.

As the level of automation has been increasing continuously in production as well as logistics, additional aspects must be considered. Although AGVs are a proven transport concept, the technical access and costs for purchase and operation have been decreasing, which contributes to their spread in transport logistics. As even handling and charging tasks are more and more fully automated and often affiliated with each other, comparing and evaluating the effects of different strategies for those tasks is usually one of the most important tasks supported by a simulation study. This requires a more detailed description of the transport vehicles and their energy attributes, as they can have a significant influence on the results. An overview of relevant parameters is given in Table 3.1.

Table 3.1 Common parameters for describing the relevant features of battery-electric vehicles in simulation models (* unit can differ, e.g., depending on simulation software)

	Parameter	Unit*	Description
Driving and handling	Driving speed empty	m/s	Driving without any load
	Driving speed loaded	m/s	Driving with loaded or hitched transport unit
	Driving speed curve	m/s	Reduced speed at intersections
	Driving speed slow	m/s	Reduced speed, e.g., for approaching station
	Acceleration/deceleration	m/s^2	E.g., general for all kinds of driving
	Rotation speed	°/s	Speed for turning on the spot
	Positioning time	s	Vernier adjustment
	Loading time	s	Load transport unit, e.g., by lifting or hitching
	Unloading time	s	Unload transport unit, e.g., by putting down or unhitching
Battery	Capacity	Ah	Overall capacity per vehicle
	Reserve	Ah	Minimum capacity, e.g., for triggering charging controls
	Charge current	A	Charge current used recharging battery
	Consumption idle	A	Basic consumption in idle state (not driving/loading/charging…)
	Consumption driving	A	Driving without any transport unit
	Consumption driving empties	A	Moving transport unit with empties
	Consumption driving material	A	Moving transport unit with material

In what manner those parameters are modeled and computed throughout the simulation depends on the used software and modeling approach. There are use cases where not all listed parameters need to be considered or can be assumed identical (also if data are not available, especially during an early planning phase). On the other hand, it might be necessary to describe driving, handling, and battery attributes even more detailed.

3.4.1.3 Example Logistics Process and Transport Management System

The modeled production in this use case consists of two parallel assembly lines with storage units alongside, where in timed operation steps front and rear doors as well as front and rear lids are taken from a mono-material container and mounted to the body frame.

As soon as a container is emptied at the assembly line, two transport demands for AGVs are reported to a transport management system:

- Empties: Driving to assembly, lifting the empty container, driving to the storage area, and placing it onto a loading station.
- Materials: Driving to the storage area, lifting a material container placed on a loading station, driving to assembly, and placing it to its defined location beneath the assembly line.

The transport within the storage area between an abstract container buffer and the loading stations is conducted by non-automated forklifts and modeled in a very abstract manner, assuming that there is always a forklift available.

The purpose of the transport management system is to administrate and rate the upcoming transport demands and to schedule tours for executing them in a prioritized order. The objective is to gain an optimal combination of demands and assigned transport vehicles as well as logistic stations, where all transports are finished within the available replenishment time. Even in systems with non-automated vehicles this is based on a complex decision framework with large sets of rules and variables. Often, this needs to relate at least the following information with each other:

- Availability of necessary transport resources, most of all vehicles fulfilling the basic needs, e.g., load limits or assignment to predefined transport routes
- Availability of necessary additional transport resources or equipment, e.g., different kinds of trailers
- Availability of free loading and unloading positions
- Replenishment time of demand
- Expected tour duration, estimated based on the current vehicle location and resulting driving distances as well as handling times

Some of the parameters listed in Table 3.1—and the computations based on them—now have to be taken into account. For each demand, several computations are necessary:

- Required amount of energy based on the expected course, distances, and sections of the tour distinguished by different consumption rates (for driving empty and driving loaded)
- Remaining battery charge of all available transport vehicles fulfilling the previous requirements

Only one of those vehicles can be assigned, where the charge will not decrease below the reserve during the tour. It always has to be possible to use the remaining charge for approaching some kind of recharging position. The reserve must be set to a suitable value depending on availability and location of those positions. Different recharging rules can be, among others:

- Always recharge to a determined battery capacity.
- Only recharge as long as vehicle is idle. If the battery capacity is sufficient for the next most urgent demand, the vehicle can be assigned to the tour.

Depending on possible more complex battery charging strategies, even more aspects become part of the decision-making of the transport management system, especially if material handling and battery charging can be combined to some extent. Several variants can be found here, differing in the required charging equipment and costs involved.

Possible kinds of strategies involving different technical equipment, e.g.,

- Stationary charging at limited charging stations (i.e., only for battery charging, while idle vehicles have to approach a free waiting station)
- Stationary charging at limited combined waiting stations with charging functionality
- Stationary charging at different kinds of stations such as loading and unloading positions with handling actions taking place in parallel
- Inductive charging during driving
- Combination of the above

If it is possible to recharge during the tour, the amount of gained battery energy can be subtracted from the energy demand of the tour.

3.4.1.4 Modeling Approach Using the "Automotive Library"

Throughout the planning, it is usually of special interest to switch between the different strategies and parameter combinations, and compare the effects. Therefore, a flexible model design is crucial for an efficient experiment phase. In this case, the simulation model is developed with the software Plant Simulation using the module "Transport Logistics" of the "Automotive Library", presented by Sokoll et al. (2021). The graphic process modeling leads to a reduction of modeling time and high flexibility as well as more intuitive understanding of the transport process and the involved decision-making. The resulting process model is shown in Fig. 3.6.

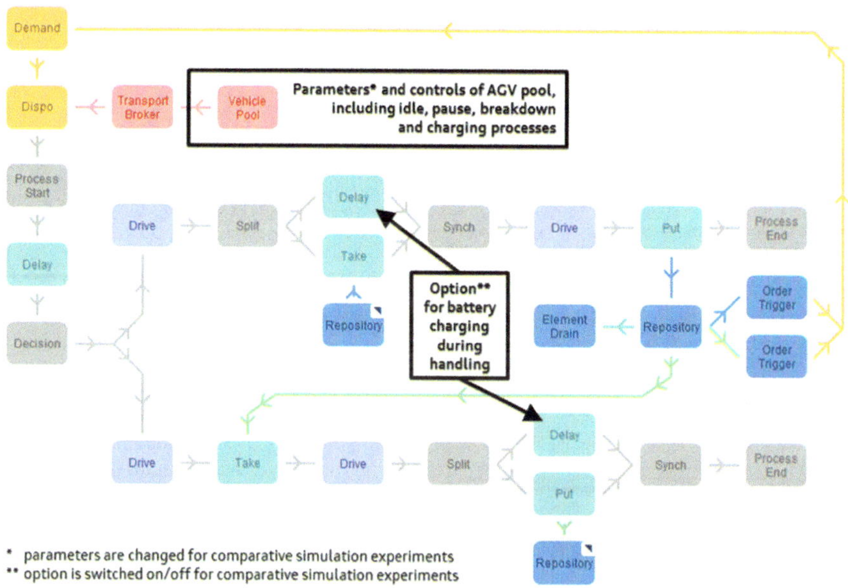

Fig. 3.6 AGV transport process modeled with the "Automotive Library" (screenshot)

All objects are a complex containers for both parameters and steering functionality. They represent either movement or handling actions, inventory and resources, or the overall demand management. They are connected with each other and steer upcoming orders by scheduling tours depending on the particularly available resources such as AGVs or logistics stations.

The previously described vehicle attributes and standard idle (driving to waiting station) as well as recharging processes (driving to charging station if battery charge falls below reserve) are located at one object within the process model called the "Vehicle Pool". In addition, an option for charging while loading or unloading a transport unit at the storage area is provided by two "Delay" objects in parallel to the according "Take" or "Put" object. This charging option does not necessarily have to be modeled in additional objects, but with these details it becomes more clear how the process is steered, and whether this option is turned on or off for an experiment. Apart from those two objects, the model is a standard AGV system with split empties and material tour. The whole process including the added charging option is executable with nearly no programming and easily modeled, assuming some amount of experience with the "Automotive Library".

In addition to the process frame, the simulation model contains a central data administration and a layout frame. Figure 3.7 shows the true-to-scale modeled layout with a network of roads and intersections, as well as the two assembly lines with containers alongside and the storage area with logistics stations for loading and unloading, waiting in the idle state, or recharging battery. Those stations are modeled in a more abstract way, as each is represented only by one station object with a given

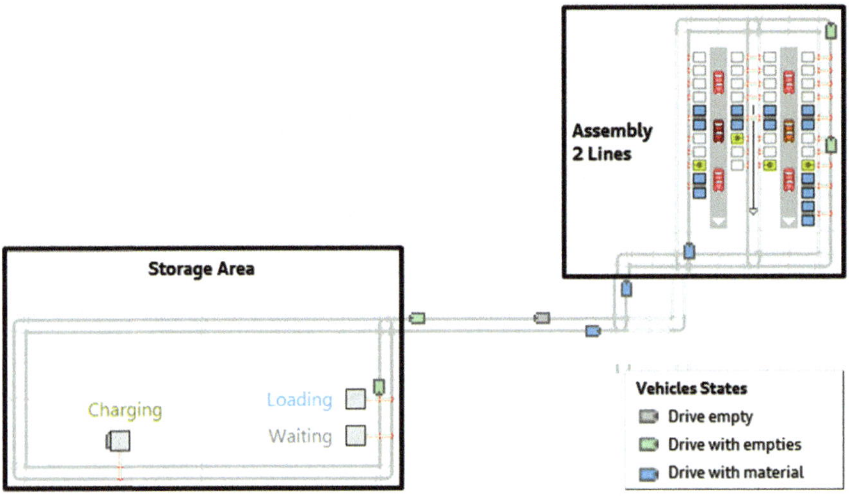

Fig. 3.7 Layout structure of the use case modeled with the "Automotive Library" (screenshot)

capacity in terms of stopping places for vehicles. Thus, it is easy to change their capacity for different simulation experiments without locating single instances for each of the *n* stations. Especially in an early planning phase, where a preliminary estimation of the required number has to be determined while the layout is subject to frequent change, this clearly reduces the modeling effort.

The layout structure, as well as assembled parts and process variants are altered, simplified and downscaled to some extent. Nevertheless, the arising questions that are to be answered with the aid of the simulation model are the same.

3.4.1.5 Results of Experiments and Evaluation

As previously mentioned, the number of possible experiments is high due to the large number of variables. Therefore, only a small selection is presented. The size of the vehicle pool is set to 14 AGVs in the first instance for all experiments and the battery reserve to 20% of the battery capacity. The capacity of the loading and waiting station are set to unlimited.

The major varied *input parameters* for the experiments are:

- The available capacity of the charging station
- The option for parallel charging during handling actions at loading stations
- Optional additional charging time in case of parallel charging

The gathered *output data* allowing us to compare the experiments include:

- The allocation of the waiting, loading, and charging station in "number of vehicles simultaneously registered" on average and maximum

- The allocation of the vehicle pool in "number of vehicles simultaneously occupied for a tour" on average and maximum
- The number of pending demands on average for showing how much open transports need to be assigned
- The tour duration from start driving to finish unloading the transport unit on average
- The duration of shortages at assembly lines if occurring during the simulation on average (one of the key figures which usually has to be reduced to zero)

These data are not sufficient for a complete evaluation of the performance and stability of the process, but deliver a first impression of the system performance and sensitivity to certain process variations. Table 3.2 gives an overview of the chosen input and Table 3.3 of the resulting output data. Figures 3.8, 3.9, 3.10, and 3.11 show the development of the vehicles' battery charge percentage from full capacity over the time span of one day for each of the four experiments described in Table 3.2.

Another output for the interpretation of the gained simulation result is the development of the vehicle battery charge over time. Figures 3.8, 3.9, 3.10 and 3.11 show the state of the batteries in detail as the development of the charge in percentage from capacity over the time span of one day. The overall simulation time was set to thirty days, but nonetheless, the graphic representation of the first day of simulation time can show the effect of different parameter and process variations in a clear way.

Evaluating Experiment 1 and Experiment 2, where vehicles have to drive to a separate charging station with limited capacity, the amount of time for charging the vehicle from reserve to capacity blocks too many resources for too much time, resulting in shortages at the assembly lines. The vehicles are fully recharged when their battery is below reserve which takes up to 30 min. The vehicles lock each other

Table 3.2 Overview of varied input

	Exp1	Exp2	Exp3	Exp4
Charging capacity	1	2	1	1
Charging at loading	False	False	True	True
Additional charging time	Not relevant	Not relevant	0.00 min	0.50 min

Table 3.3 Overview of compared output

	Exp1	Exp2	Exp3	Exp4
Waiting allocation	0.36 [max. 14]	0.60 [max. 14]	1.32 [max. 14]	1.35 [max. 14]
Loading allocation	0.89 [max. 6]	1.42 [max. 6]	1.42 [max. 6]	1.70 [max. 7]
Charging allocation	0.39 [max. 1]	1.19 [max. 2]	0.17 [max. 1]	0.00 [max. 0]
Vehicle reservation	13.66 [97.57%]	13.45 [96.04%]	12.78 [91.32%]	12.75 [91.06%]
Pending demands	44.95 [max. 71]	17.42 [max. 39.2]	13.39 [max. 23]	13.42 [max. 23]
Tour duration	19.55 min	19.53 min	19.65 min	19.97 min
Shortage duration	50.5 min	10.88 min	0.00 min	0.00 min

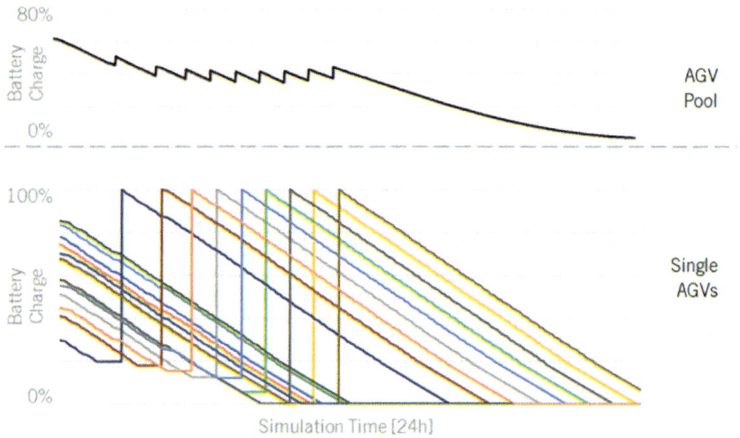

Fig. 3.8 Development of battery charge in Experiment 1: single charging capacity, no charging on load

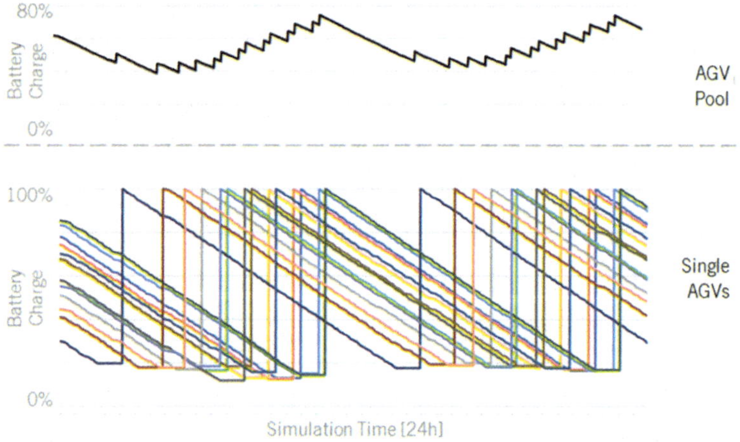

Fig. 3.9 Development of battery charge in Experiment 2: double charging capacity

at the charging station and one by another is running out of energy while waiting for a free charging slot. Further experiments show that even up to four charging stations are not sufficient. The reason is that the recharging time is too long compared to the replenishment times of the transport demands. The development of the battery state over time in Fig. 3.8 also shows that the reserve is not high enough to cover the delay until the charging can be started: It drops to zero for all AGVs during the simulation time of one day.

Doubling the number of charging stations clearly decreases the number of pending demands and shortage duration, as vehicles spend less time waiting for a free slot and

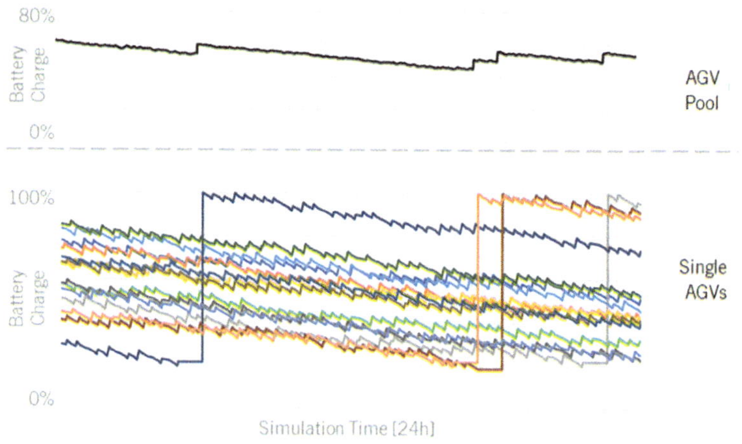

Fig. 3.10 Development of battery charge in Experiment 3: charging on load, no additional charging time

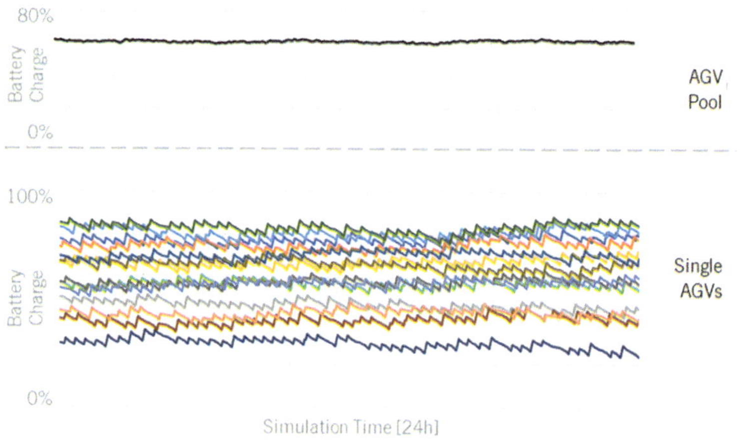

Fig. 3.11 Development of battery charge in Experiment 4: charging on load, 30 s additional charging time

no AGV breaks down due to an empty battery. Nevertheless, a further optimization approach is necessary.

Therefore, in Experiment 3 and Experiment 4, the option for parallel loading and charging is switched on, having a significantly positive effect on the results. No shortages occur during the simulation run and vehicles increase the charge in discrete steps while finishing the handling of the transport unit at the storage area.

But, the overall trend of the battery charge in Experiment 3 is still a negative one, showing that the gained energy is less than the energy demand for the rest of the tour. It happens that after a certain amount of time the battery needs to be fully recharged at the separate charging station. Based on that result, experiments were conducted to what extent the charging time has to be prolonged at the loading station to balance energy recharges and demands. In intermediate steps, the required time has been determined and finally set to 0.50 min in Experiment 4. As a result, the development of the battery charge over time of all AGVs is stable, but the allocation of the loading station increases as it is not limited. Further experiments prove that a limitation of the capacity down to two stations, based on the previously determined average allocation, would be sufficient. As the allocation of the vehicles by the recharging process at the charging station decreases from experiment to experiment, the number of AGVs required for a sufficient supply also decreases.

However, this cannot be the final recommendation and many other parameters may have a more significant influence and, thus, be related to the overall set of input data. Examples might be different technical availabilities of the charging equipment or the available charging power. Assuming that high power stations recharge batteries faster, their use would result in a decrease of recharging times, and presumably in a lower number of stations required. On the other hand, those kinds of stations are associated with higher costs but might also differ in their availability. As an example, at the beginning of the project, it was intended to provide each material station at the assembly lines with a charging functionality. Later simulation experiments proved that fewer charging options per line are sufficient. Assuming investment costs of around 2,000 Euro per charging plate that need to be installed at a material station, the reduction of costs only for this aspect has reduced the investment by a lower six-figure amount in the real system.

The results in Table 3.2 and Fig. 3.8 are only a small selection from a very broad number of experiments, simulated during the overall planning project, which included more than twelve variations of process models (compared to Fig. 3.6) and more than 100 experiment settings. The interpretation and evaluation of the gained simulation results involve many factors that need expertise covering all procedural, technical and economic aspects.

3.4.2 Energy Simulation in an Aluminum Foundry

In consequence of climate change and energy transition, the importance of energy efficiency has increased sharply, especially in Germany. In the automotive industry, too, the growing environmental awareness of the population and political requirements are highlighting the CO_2 balance and the energy efficiency of production and products. The potential for energy and cost savings is particularly high in energy-intensive industries as described in Sect. 3.1, such as the aluminum die casting industry, which is a supplier for the automotive sector. In this industry, energy consumption per ton of good casting ranges between 2,000 and 6,000 kWh (Belt 2015; Bosse et al. 2013; Herrmann et al. 2013). According to the German Federal Statistical Office, this leads to a high energy cost burden, which can exceed 25% of the gross value added (Bundesverband Deutscher Gießereien, 2022). Up to 60%

of the energy is used for fusing the aluminum mostly in gas-fired melting furnaces. Due to the increasing volatility of gas prices, the economic pressure on companies is growing. Therefore, the simulation-based analysis of options for more energy-efficient production operations is of outstanding importance. In aluminum foundries, the energy efficiency of the used melting furnaces depends highly on their design, age, utilisation, and mode of operation (Belt 2015; Felder et al. 2014; Stephan et al. 2005; Salonitis et al. 2016). To study energy-efficient control of the production processes in aluminum foundries, simulation-based methods are used (Fuss et al. 2013; Krause et al. 2012; Herrmann et al. 2011). The studies show energy efficiency potentials for the respective operating components. However, the small plant sizes considered in these studies do not allow for developing control strategies for efficient melting furnace operation in a larger die casting plant.

3.4.2.1 Initial Situation and Goals of the Study

The focus of the study is a foundry that supplies the automotive industry. Forklifts are deployed for the transport within the production area. A major goal of the study was to determine the impact of downtime of the die casting machines on the total productivity.

The plant structure of the melting area to be represented in the simulation contains four melting furnaces with a total melting capacity of 11.3 t/h. The casting area consists of 31 casting machines, with eleven casting machines (large parts) having a high shot weight of 25 kg and a long cycle time of 120 s. The remaining 20 casting machines (small parts) are operated with a smaller shot weight of 5 kg and a shorter cycle time of 80 s.

3.4.2.2 Model Description

In the first step, the production structure of a typical aluminum die casting plant was analyzed and displayed in a diagram (Fig. 3.12). The underlying processes to be simulated are:

- Delivery of liquid aluminum (Fig. 3.12a),
- Charging of the gas-fired shaft melting furnaces (MF) via forklifts with ingots (metal bars), return or scrap material (Fig. 3.12b)
- Heating, melting, and superheating the aluminum or keeping the metal warm (Fig. 3.12c)
- Distribution of the molten aluminum with forklifts to the metering furnaces of the die casting machines (Fig. 3.12d)
- Production of castings in the die casting machines (DCM) and quality inspection (Fig. 3.12e)
- Transport of material containers from the die casting store or ingot packages from the warehouse to the melting store (Fig. 3.12f)

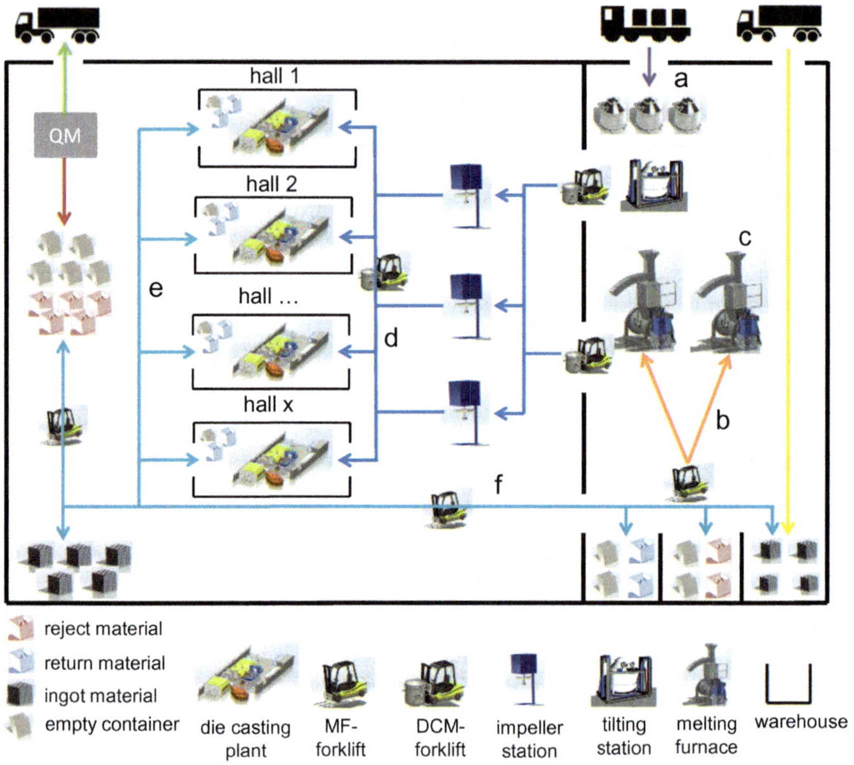

Fig. 3.12 Plant components and process steps of an aluminum die casting plant

Three forklifts are used to distribute the liquid aluminum to the various die casting machines (process step d), and three forklifts supply the melting furnaces with solid aluminum (process step f). Due to the large number of identical plant components, which have different process parameters, an object-oriented structure of the simulation is necessary. Each plant component from Fig. 3.12 is modeled by a corresponding class. To simplify the simulation of individual operating scenarios, the parameterization of the individual objects and the operating data is conducted via a table.

The core of the targeted simulation is the energetic model of the melting furnaces. The melting and holding processes (process step c) represent complex thermodynamic processes that are influenced by many factors. Based on the mathematical modeling of the thermodynamic processes in the melting furnace, the energy model was implemented as a block diagram.

The process control module determines the control interventions for the operation of the melting furnaces and the forklifts. For this purpose, the plant and process parameters of the material and energy flow model are loaded and evaluated. From these data, suitable control interventions are derived, which are transferred from the process control to the coupled energy and material simulation. The development of the process control proved to be very extensive, because different control strategies were implemented for reasons of comparison.

3.4.2.3 Modeling Application Aspects

The consideration of energy aspects with the desired detailed mapping of the melting and holding process in the melting store requires a hybrid simulation. The level of detail goes significantly beyond the mere allocation of energy consumption to the individual processes. The hybrid simulation of the various processes has been implemented by a combination of the MATLAB, Simulink, and Stateflow simulation tools The distribution of tasks has been defined as follows:

- MATLAB: simulation control, object instantiation and management, evaluation.
- Simulink: simulation of continuous processes
- Stateflow: simulation of discrete event processes

The realization of material flow, energy model, and control in a single software environment eliminates the need for complex and time-consuming coupling of different software packages for the discrete event material flow model and the continuous energy model.

The complete simulation model, consisting of the submodels energy flow, material flow, and process control, is shown in Fig. 3.13. The energy flow model is used to calculate the thermodynamic processes of the aluminum melting furnaces. The material flow model serves for recording the complete material flow within the plant. There is a bidirectional coupling between the material flow model and the energy flow model: on the one hand, the energy model is provided with information on the input and output of solid or liquid aluminum (Fig. 3.13a). On the other hand, the energy model transfers the most important furnace data to the material flow model, such as the current melting rate or the amount of available liquid aluminum (Fig. 3.13b).

In the process control (Fig. 3.13c), different control strategies can be implemented and compared with respect to their effects on production and energy consumption. In the simulation, the production orders for the die casting machines are assigned in addition to switching on and off (simulation of downtimes) the components. The operation-dependent order data determination compiles possible orders "just in time" and identifies the optimal order. The control system is used to manage the storage of the solid aluminum and the charging of the melting furnaces with solid material. Essentially, the time, mass, and material type of the charge are specified. The control system is, thus, decisive for the correct filling level of the solid aluminum in the melting shaft. Therefore, the simulation model can investigate and analyze the effects of different forklift control concepts on the melting furnaces.

A reliable supply of aluminum to the die casting machines is crucial for production operations. Therefore, the control of the forklifts of the DCM production for the distribution of liquid aluminum from the melting furnaces via impeller station to the die casting machines is of prime importance. The control algorithm determines

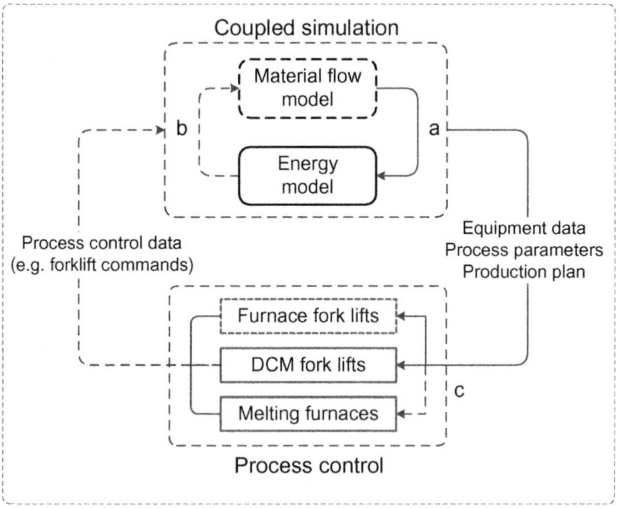

Fig. 3.13 Structure of the simulation model

the material source (melting furnace), the material sinks to be filled (die casting machines) and the resulting withdrawal quantity. The die casting machines to be filled can be selected based on various criteria. In industry, the selection is usually based on a traffic light system with a defined signal color (red—yellow—green) for a specific filling level range. This selection mechanism can be improved by recording and comparing the exact machine fill levels. Another alternative is a distribution based on the remaining runtimes of the die casting machines. In this case, for each metering furnace the remaining runtime until a shutdown due to material shortage must be calculated, according to the current filling level, the given cycle time, and the product-dependent shot weight. The metering furnace that has the shortest remaining runtime is filled first. This is the optimal strategy.

The basic physical model of the melting furnaces is based on a dynamic system of differential equations. Starting from a combustion calculation, it computes the heat transfer from the flue gas to the aluminum and simulate the melting and holding process in the melting furnace. The material and energy flow in the furnace is bidirectionally coupled in the simulation model (Fig. 3.14). The model includes the changing of the solid aluminum surface as well as the heat loss via the furnace wall. Both factors influence the heat transfer from flue gas to aluminum.

For each component from Fig. 3.12 (melting furnace, die casting machine, forklift, etc.), a state chart was defined in Stateflow. In the state charts, the different states of the components are connected by event-driven state transitions. To enable communication between the individual systems, an interface between the Simulink and Stateflow models was required. This has been realized by "Interpreted MATLAB Functions" (Fig. 3.15.). The IN function passes on the necessary system parameters and events to the corresponding components. In each time step, the OUT function

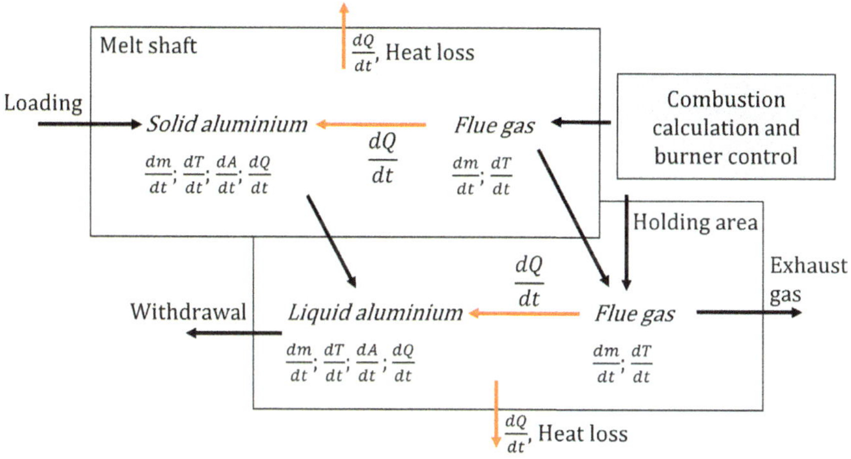

Fig. 3.14 Schematic representation of the melting furnace model

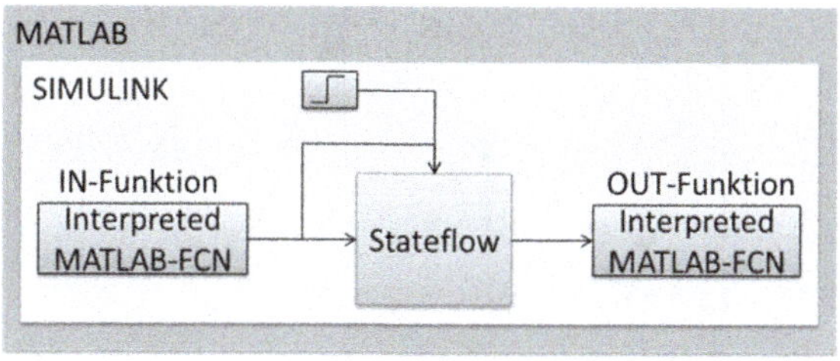

Fig. 3.15 Interface between MATLAB, Simulink, and Stateflow

determines the value changes of the system by querying the status of the component. Here, the adaptation of the process parameters takes place depending on the status.

In industrial practice, control strategies that are difficult to describe algorithmically often occur. With the line-oriented implementation of the sequence control in MATLAB, these can be implemented relatively easily. Stochastic influencing variables of order determination and execution can be taken into account in the programming structure.

3.4.2.4 Results of Experiments and Evaluation

To validate the simulation, data acquisition was carried out in a large aluminum die casting plant, which has good data availability of process data. In addition to determining the feeding strategy and the production schedule, the operating data were recorded at all shifts of a complete calendar week. For the material flow model, the subsequent simulation revealed deviations in the number of aluminum parts produced of 1.4% and the amount of aluminum consumed of 0.9%. These are due to unplanned shutdowns of the die casting machines, which can only be statistically approximated but not precisely predicted.

Based on the recorded data, the correctness of the energy model was also confirmed. The molten aluminum mass and the gas consumption deviate from the actual values by 1.5% and 0.5%, respectively. The accuracy of the thermodynamic calculations can be seen not only by the molten aluminum mass (Fig. 3.16), but also by the time curve of the flue gas temperature (Fig. 3.17).

The validation results demonstrate that the simulation model can be used for representing real operations accurately. Based on the initial configuration, various operating parameters can be varied and the impact on material and energy flow can be recorded. These include:

- Preheating of the solid aluminum by adjusting the ingot temperature
- Variation of the forklift control (feeding strategy)
- Variation of the externally delivered liquid aluminum quantity (Fig. 3.12a)

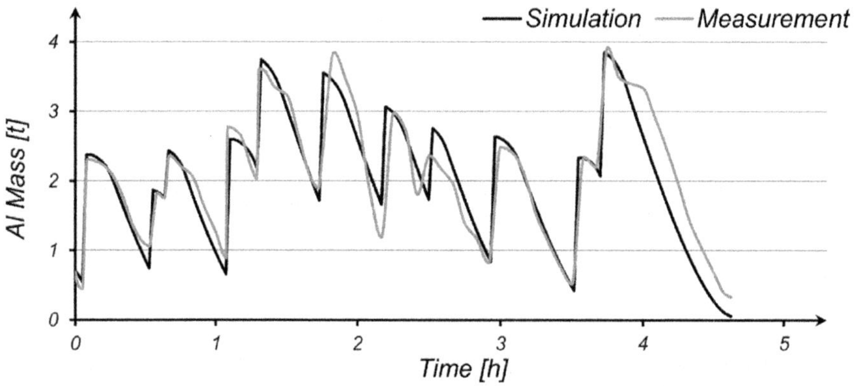

Fig. 3.16 Validation of the energy model based on the molten aluminum mass

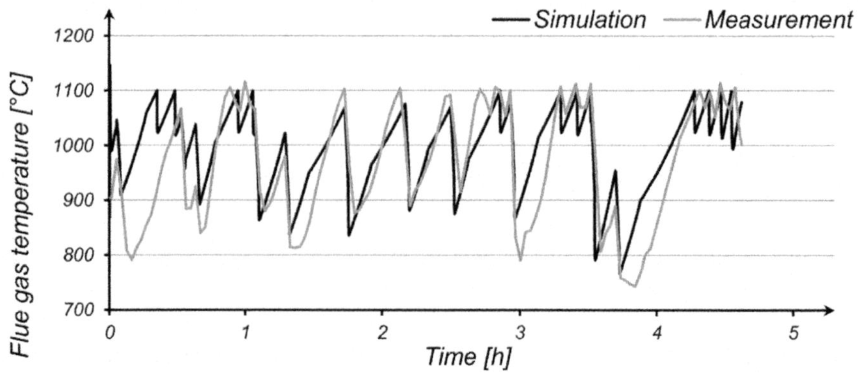

Fig. 3.17 Validation of the energy model based on the gas temperature

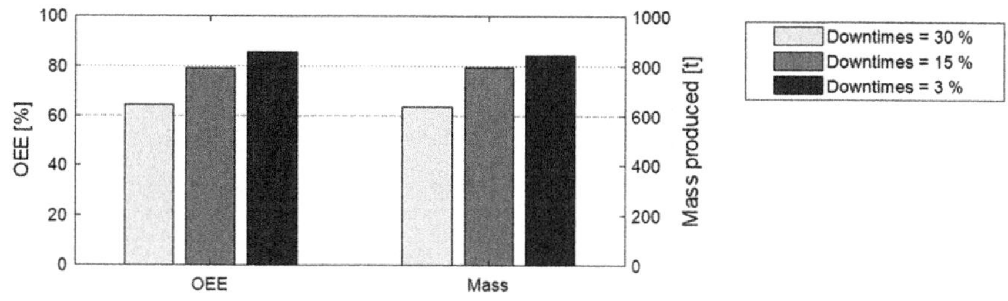

Fig. 3.18 Impact of downtime on productivity

- Variation of the liquid aluminum consumption by adjusting the downtimes
- Variation of liquid aluminum production by shutting down individual furnaces

This allows for evaluating individual efficiency measures according to the relevant factors of productivity, energy efficiency, and production reliability. In this context, production reliability means the sufficient supply of liquid aluminum to the producing die casting machines. The Overall Equipment Effectiveness (OEE) of die casting machines is a well-known measure of system efficiency based on the product of quality, performance, and availability. In addition to the number of aluminum parts produced, the OEE is particularly suitable as an evaluation figure for the productivity of the operation.

In the measurements of the reference operation examined, downtimes (planned and unplanned) of the die casting machines amount to 30% of the total running time. In this configuration, the achievable productivity of the operation is only limited by the die casting machines. Therefore, there is a surplus of liquid aluminum. By reducing downtime (e.g., by implementing an improved maintenance concept), the productivity of the overall operation can be significantly increased. The effects of reduced downtimes on productivity are given in Fig. 3.18.

The OEE can be increased from 64 to 79% because of a downtime reduction to 15% of the total running time. A further reduction of the downtimes to 3% again results in a noticeable increase of the key figure to 85%, whereby a weakening of the effect is noticeable due to the onset of the liquid aluminum shortage.

Figure 3.18 also shows that the aluminum mass produced by the melting furnaces increases in direct proportion to the availability of die casting machines. As a side effect of the increase in productivity, there is an increase in the utilization of the melting furnaces. For high productivity values (downtimes 3%), however, the liquid aluminum consumption can exceed the capacity of the melting furnaces. In addition to the planned and unplanned downtimes considered so far, die casting machine failures can also occur due to a lack of aluminum.

To investigate these efficiency measures, one smelter (out of four in the reference operation) is switched off within the simulation. The resulting specific energy consumption of the remaining furnaces is compared with the values occurring in real operation (Fig. 3.19). In a further simulation run, the liquid aluminum supply is

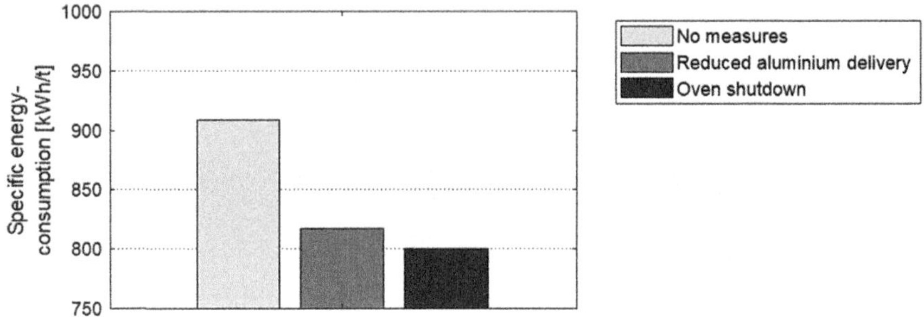

Fig. 3.19 Specific energy consumption for various savings measures with downtimes of 30%

significantly reduced (10% of the original value). With downtimes of 30%, no operational failures occur due to aluminum shortages, so production reliability remains guaranteed.

With the help of the measures described, a reduction in specific energy consumption of 10% (reduced liquid aluminum delivery) or 12% (furnace shutdown) can be realized. Unlike the preheating of solid aluminum, however, the reduction of the production capacity for liquid aluminum is not associated with an increase in production reliability, but even represents a risk. Therefore, corresponding measures should only be implemented after detailed preparation. By simulation, it is possible to determine improvement potentials without endangering real operations.

3.4.2.5 Benefits and Lessons Learned

Programming in the MATLAB/Simulink/Stateflow software package is very complex compared to the use of simulation software for discrete event simulation (such as Plant Simulation) in combination with a dynamic simulation of the thermodynamic processes. However, it offers the opportunity of an ongoing bidirectional coupling of the material and energy flow. An object-oriented approach is necessary to master the software complexity. In addition, excellent knowledge of the interaction of the MATLAB programming language with the block-oriented simulation environment Simulink and the event-oriented simulation environment Stateflow is required.

In order to achieve a comparable range of functions with special software for the simulation of production systems, the chosen approach has the following advantages:

- Detailed investigation of specific process parameters
- Simple bidirectional coupling of discrete event and continuous dynamic processes
- Simple implementation of complex control algorithms

However, this is countered by:

- The considerable time required to implement the simulation
- The necessary know-how for a very specific development environment
- No real time animation of the simulation

The parameterization of the individual components in several linked tables enables machine parameters to be changed easily, which made it possible to simulate further aluminum die casting operations. The simulation was successfully validated a second time through the operating data of a medium-sized die casting operation. With the dynamic setup of the plant structure in Simulink before each simulation run, any aluminum die casting plant whose operational structure corresponds to that in Fig. 3.10 can be simulated. The results of energy efficiency measures in aluminum die casting plants of different sizes can be displayed in an app (Hochschule Ansbach 2022). Gas savings, system efficiency, and the production balance are computed for selected company sizes and specific measures. The combination of efficiency measures shows a specific energy saving in kWh/t_{AL} of up to 24.6%.

In real operation, the simulated effects of the energy efficiency measures can only be partially achieved due to established operational processes or thermal losses. However, the results can be used to weigh up the order of magnitude of the energy savings achieved by the individual measures. For example, it can be deduced that a logistical measure in form of optimized charging processes compared to a cost-intensive preheating chamber for preheating the solid aluminum offers improvements in productivity, production reliability, and energy efficiency. The simulation also depicts that the above-mentioned measures can only realize their full potential in combination with significant reductions in plant downtimes or with the installation of additional die casting machines.

3.5 Conclusions and Outlook

The method of simulation is already widespread in the automotive industry. Although energy-related simulation is not comparably common, the pioneering role of the automotive industry is also evident here as most of the identified research publications describe use cases from the automotive industry. The diversity and complexity of this topic is reflected in the many different approaches and modeling techniques. While simulation is now widespread in production and logistics due to the increased complexity of the processes, energy-related simulation has yet to establish itself in practice. This can only succeed if economic, ecological, and political advantages can be achieved through the simulation. In order to reduce the effort involved in modeling and simulation, further standardization is necessary.

References

Alvandi S, Bienert G, Li W, Kara S (2015) Hierarchical modelling of complex material and energy flow in manufacturing systems. Procedia CIRP 29:92–97

Beier J, Neef B, Thiede S, Herrmann C (2016) Integrating on-site renewable electricity generation into a manufacturing system with intermittent battery storage from electric vehicles. Procedia CIRP 48:483–488

Belt C (2015) Current state of aluminum melting and holding furnaces in industry. J Miner Metals Mater Soc 67(11):2690–2695

Bosse M, Frost E, Hazrat M, Rhiemeier JM, Wolff H (2013) Ermittlung von branchenspezifischen Potentialen zum Einsatz von erneuerbaren Energien in besonders energieintensiven Industriesektoren am Beispiel der Gießerei- Industrie. IfG Institut für Gießereitechnik. Ecofys Germany, Düsseldorf

Brüggemann H, Geckler D, Laumeyer M (2014) Energieeffizienz bei Industrierobotern. VDI-Z 6:55–57

Bundesverband Deutscher Gießereien (2022) Referat Energie und Umwelt. https://www.guss.de/organisation/bdg/referat-energie-und-umwelt. Accessed 22 Feb 2022

Chu KC, Kaifuku K, Saitou K (2016) Optimal integration of alternative energy sources in production systems with customer demand forecast. IEEE Trans Autom Sci Eng 13(1):206–214

Dunkelberg H, Heidrich T, Weiß T, Hesselbach J (2020) Energy demand flexibilization of industrial consumers. J Simul 14(1):53–63

Felder P, Flückiger U, Wisskopf T (2014) Gießereibetrieb – Pinch-Analyse Prozessanlagen und Heizungsanlagen. Schlussbericht. EnergieSchweiz, Bern

Fuss E, Drvendzija S, Krewald A (2013) Transparenz als Schlüssel zur energieeffizienten Produktion: Ablaufplanung in Schmelzerei und Gießerei mittels planungsbegleitender Energie- und Materialflusssimulation. Product Manage 18(3):35–38

Geckler D, Fliege H, Nagel J, Bracht U, Seewaldt M, Wolff D (2020) Der Energiebedarf der Automobilproduktion im Fokus der Simulation. In: Mayer G, Pöge C, Spieckermann S, Wenzel S (eds) Ablaufsimulation in der Automobilindustrie. Springer, Berlin Heidelberg, pp 303–317

Herrmann C, Thiede S, Kara S, Hesselbach J (2011) Energy oriented simulation of manufacturing systems—concept and application. CIRP Ann 60(1):45–48

Herrmann C, Pries H, Hartmann G (2013) Energie- und ressourceneffiziente Produktion von Aluminumdruckguss. Springer Vieweg, Berlin

Hochschule Ansbach (2022) E|Melt Demonstratorsystem. https://www.hs-ansbach.de/emelt. Accessed 22 Feb 2022

Ichimura H, Takakuwa S (2013) Decision making on manufacturing system from the perspective of material flow cost accounting. In: Pasupathy R, Kim SH, Tolk A, Hill RR, Kuhl ME (eds) Proceedings of the 2013 Winter Simulation Conference. IEEE, Piscataway, NJ, pp 1973–1983

Krause M, Thiede S, Herrmann C, Butz F (2012) Material and energy flow oriented method for enhancing energy and resource efficiency in aluminum foundries. In: Dornfeld D, Linke B (eds) Leveraging technology for a sustainable world. Springer, Berlin Heidelberg, pp 281–286

Kuhlmann T, Sauer A (2019) Design and assessment of energetic agility measures in factories based on multivariate linear regression. Procedia CIRP 80:162–167

Lee JY, Kang HS, Noh SD (2012) Simulation-based analysis for sustainability of manufacturing system. Int J Precis Eng Manuf 13(7):1221–1230

Mayer G, Pöge C, Spieckermann S, Wenzel S (2020) Einleitung. In: Mayer G, Pöge C, Spieckermann S, Wenzel S (eds) Ablaufsimulation in der Automobilindustrie. Springer, Berlin Heidelberg, pp 1–17

Neugebauer R, Putz M, Schlegel A, Langer T, Franz E, Lorenz S (2012) Energy-sensitive production control in mixed model manufacturing processes. In: Dornfeld DA, Linke BS (eds) Leveraging technology for a sustainable world. Springer, Berlin Heidelberg, pp 399–404

Omar MA, Qilun Z, Lujia F, Ali AA, Lahjouji D, Khraisheh M (2016) A hybrid simulation approach for predicting energy flows in production lines. Int J Sustain Eng 9(1):25–34

Poeting M, Prell B, Rabe M, Uhlig T, Wenzel S (2019) Considering energy-related factors in the simulation of logistics systems. In: Mustafee N, Bae KHG, Lazarova-Molnar S, Rabe M, Szabo C, Haas P, Son YJ JW (eds) Proceedings of the 2019 Winter Simulation Conference. IEEE, Piscataway, NJ, pp 1849–1858

Salonitis K, Zeng B, Mehrabi HA, Jolly M (2016) The challenges for energy efficient casting processes. Procedia CIRP 40:24–29

Sinnemann J, Bartelt M, Strahilov A, Kuhlenkötter B (2020) Architecture for simulation and optimization of energy consumption of automated production systems. Procedia CIRP 93:1241–1246

Sivapragasam E (2016) Entwicklung eines Ansatzes zur digitalen Planung und Optimierung von Energieverbräuchen am Beispiel der Druckluftbereitstellung. Beuth Hochschule für Technik, Berlin

Sokoll K, von Braunschweig P, Gustafsson-Ende L, Hilmer F (2021) Ein unternehmensübergreifender Standard für Logistiksimulation mit grafischer Prozessmodellierung. In: Franke J, Schuderer P (eds) Simulation in Produktion und Logistik 2021. Cuvillier, Göttingen Germany, pp 505–514

Stange C, Bös M (2019) Simulationsgestützte Lernumgebung für Energieeffizienz. In: Putz M, Schlegel A (eds) Simulation in Produktion und Logistik 2019. Verlag Wissenschaftliche Skripten, Auerbach, pp 213–222

Stephan W, Zitzmann K, Pröbstle G, Kapischke J (2005) Effiziente Energieverwendung in der Industrie, Bayerisches Landesamt für Umweltschutz. Augsburg

Thiemicke S (2016) Simulative Potentialanalyse von Energieeffizienzmaßnahmen in der Volkswagen Komponentenfertigung. Master thesis, Technische Universität Braunschweig, Braunschweig

Thomitzek M, von Drachenfels N, Cerdas F, Herrmann C, Thiede S (2019) Simulation-based assessment of the energy demand in battery cell manufacturing. Procedia CIRP 80:126–131

Tur B, Fischer E, Franke J, Bergs CS, Schulze C (2019) Modelling of energy storage devices and converters for energy flow simulation in Plant Simulation. In: Putz M, Schlegel A (eds) Simulation in Produktion und Logistik 2019. Verlag Wissenschaftliche Skripten, Auerbach, pp 97–109

Visnic B (2020) Tesla casts a new strategy for lightweight structures. https://www.sae.org/news/2020/06/tesla-model-y-big-castings. Accessed 22 Feb 2022

Wenzel S, Halfar F, Pöge C, Spieckermann S (2015) Simulationsgestützte Planung energieeffizienter Produktionssysteme. ZWF 110(7–8):403–406

Wenzel S, Peter T, Stoldt J, Schlegel A, Groß G, Pitsch H, Rabe M, Seewaldt M (2017) Betrachtungen energetischer Einflussfaktoren in der Simulation in Produktion und Logistik: Eine Literaturanalyse. In: Wenzel S, Peter T (eds) Simulation in Produktion und Logistik 2017. Kassel University Press, Kassel, pp 9–18

Wilson J, Arokiam A, Belaidi H, Ladbrook J (2016) A simple energy usage toolkit from manufacturing simulation data. J Clean Prod 122:266–276

Chapter 4
Transportation

Marvin Auf der Landwehr, Javier Faulin, and Adrian Serrano-Hernandez

Abstract With the evolution of emerging technologies, transportation systems are becoming increasingly complex. At the same time, the advent of grand environmental challenges, such as climate change, requires researchers and practitioners to develop new transportation strategies that ensure a high degree of energy efficiency. Due to their immanent capabilities to study the behavior of complex systems over time, simulation methodologies can provide valuable assets to determine the energy efficiency on a transportation system. Thus, this chapter reviews the current state of the art regarding energy-related simulation research in the transportation sector. It outlines the status quo of energy-related simulation research and provides an overview on the most common simulation methods used for analyzing energy-related transportation aspects such as vehicle emissions. Moreover, to demonstrate the practical applicability of simulation in this domain, two exemplary use cases are elaborated, employing an agent-based modeling technique to assess emission implications resulting from different freight and grocery transportation strategies. The results of the use cases show that simulation can be a powerful methodology to evaluate energy-related transportation effects, ultimately supporting more informed theory construction and strategy formulation incentives.

4.1 Introduction

With the rising importance of human capital, technology and process efficiency for industrial competitiveness, and the growing demands on prices and product quality, companies across the globe are forced to continuously reconfigure their product portfolio, their operating methods, their market approaches, as well as their processes

M. Auf der Landwehr (✉)
Fakultät Wirtschaft und Informatik, Institute for Applied Data Science Hanover, Hochschule Hannover, Ricklinger Stadtweg 120, 30459 Hannover, Germany
e-mail: marvin.auf-der-landwehr@hs-hannover.de

J. Faulin · A. Serrano-Hernandez
Institute of Smart Cities, Department of Statistics, Computer Science, and Mathematics, Public University of Navarra, Campus de Arrosadia, 31006 Pamplona, Spain

S. Wenzel et al. (eds.), *Energy-Related Material Flow Simulation in Production and Logistics*, https://doi.org/10.1007/978-3-031-34218-9_4

for collecting materials and distributing products. To this respect, freight transportation is a key determinant in business, as it allows for transferring goods from places where they are produced to places where they are required. Thus, transportation is a fundamental service that is indispensable for supply chains linking the organization to suppliers and to customers.

At the same time, transport is a major source of environmental pollution and contributes to climate change, air pollution, and noise emissions, among other externalities (Demir et al. 2015). According to the European Environment Agency (2021a), transportation consumes a third of all energy sources in the European Union (EU). Most of this energy derives from oil, although renewable energies are gaining a more important role in energy production (Faulin et al. 2006). Transportation is responsible for about 26% of the EU's carbon dioxide (CO_2) emissions and contributes significantly to climate change (European Environment Agency 2021b). While most of the other economic sectors, for example, industry and energy production, have reduced their emissions since the 1990ies, emissions from transport have increased. Currently, these represent more than a quarter of the total greenhouse gas emissions in the EU (European Environment Agency 2021a).

Transportation systems are known to be complex. They are usually characterized by a large number of elements (e.g., all actors in a supply chain model) and by highly uncertain interrelations among them. Additionally, this complexity is not an inherent property of the behavior of the system itself, but rather a result of lacking methodologies and tools that could enable the development of a model that is capable of reliably mimicking the system under study. The paradigm of last mile distribution is a good example for the complexity of transportation systems (Janjevic and Winkenbach 2020; Melkonyan et al. 2020). Broadly speaking, there are multiple conflicting objectives among different stakeholders in key delivery activities in city centers that lead to both, opportunities and challenges for sustainable cities.

In particular, three problems arise from last-mile distribution (Olsson et al. 2019; Melkoyan et al. 2020; Boysen et al. 2021):

1. Last mile distribution represents around a third of all transportation *costs* in an extremely competitive and crowded sector, which crushes profitability.
2. Small trucks and vans are responsible for a significant part of traffic *congestion* in cities.
3. Transportation is a major source of *air pollution*, especially in cities. Apart from CO_2 emissions, air pollutants, such as nitrogen dioxide (NO_2) and particulate matter (PM), are particularly harmful to human health and the environment.

Major stakeholders for the two main delivery domains, namely e-commerce and hospitality, are consumers, logistics service providers (LSP), and local administrations.

Potential solutions or enablers may build on infrastructural measures and inter-organizational cooperation. On the one hand, an appropriate network of automated parcel lockers, in-city consolidation centers, the optimization of load and unload zones, among other strategies, could improve the freight flows. This would also encompass a push towards changing vehicle technologies, i.e., electric vans or cargo bikes. On the other hand, horizontal cooperation among LSPs could optimize load capacities and reduce the total number of vehicles required for a delivery operation. Thereby, cooperation is not limited to companies. Instead, public–private cooperation is also a requirement to boost sustainable last mile distributions, even though it would require to align regulations from local, regional and national administrations.

In order to measure, track, evaluate, and optimize the energy-related implications of complex transportation systems, a dedicated methodology is required that is able to holistically capture their behavior, even if it is fairly unknown. This comes particularly true, when environmental aspects have to be taken into consideration, entailing the inclusion of energy-related variables such as fuel consumption or CO_2 emissions. Simulation can constitute the key for this development. To position simulation as a solution methodology in the transportation domain, this chapter reviews simulation methodologies for energy-related transportation aspects and reports on two simulation-based case studies for last-mile distribution. Finally, remarks on the insights and specifics of this chapter are summarized.

4.2 Energy-Related Transportation Simulation

Scientific literature is full of examples employing simulation models to cope with complexity in transportation systems with a special focus on energy-related aspects. Corlu et al. (2020) provide a general overview of modeling energy in transportation. In their literature review, several methodologies are elaborated on, concluding that the algorithms that combine metaheuristics with simulation can provide the required flexibility for dealing with real-world dynamic scenarios. From an illustrative point of view, Fig. 4.1 outlines the number of Scopus-index articles published in the period from 1970 to 2020 covering the topic of energy-related simulation in transportation (search string: simulation *and* transportation *and* (fuel and consumption *or* energy *or* emissions)). In contrast, Fig. 4.2 exposes the most prominent journals on this topic from 2010 to 2020. In brief, there is a remarkable and nearly exponential increase from 25 articles in 2005 up to 240 publications published in 2020. These articles are mainly published in transportation- and energy-related journals with international target audience, including *Transportation Research Part D: Transport and Environment*, *International Journal of Hydrogen Energy*, and *Applied Energy* as the most prominent scientific platforms.

The work of Crainic et al. (2018) constitutes a starting point. This article reviews intermodal freight transportation from the perspective of simulation methodologies. Crainic et al. concluded that urban contexts and ecological implications have experienced a significant growth in research interest, with most analyses focusing on the reduction of greenhouse gas emissions as the ultimate goal. Thereby, environmental factors are mainly considered by means of generalized costs in urban areas, either directly or indirectly. Hence, for example, the reduction in traffic, road occupancy, and kilometers traveled are often measured in carbon dioxide equivalent units. Only a few publications directly estimate greenhouse gas emissions by measuring CO_2 or

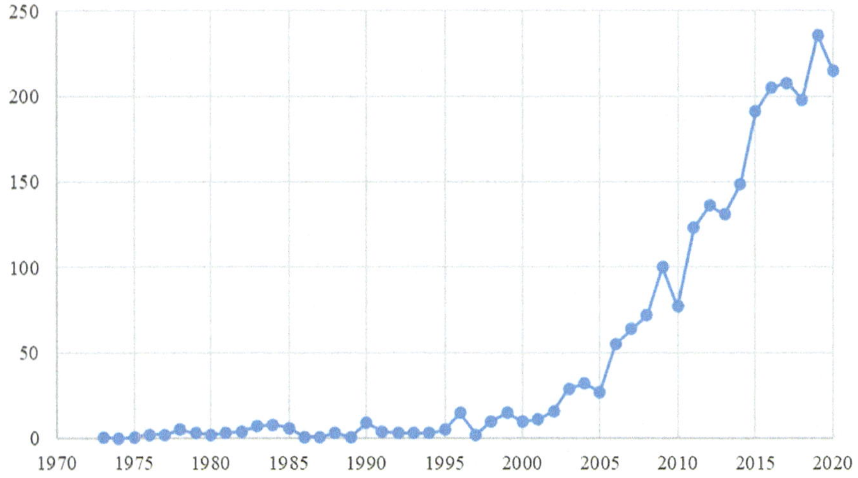

Fig. 4.1 Scopus-index articles of energy-related simulation in transportation since 1970

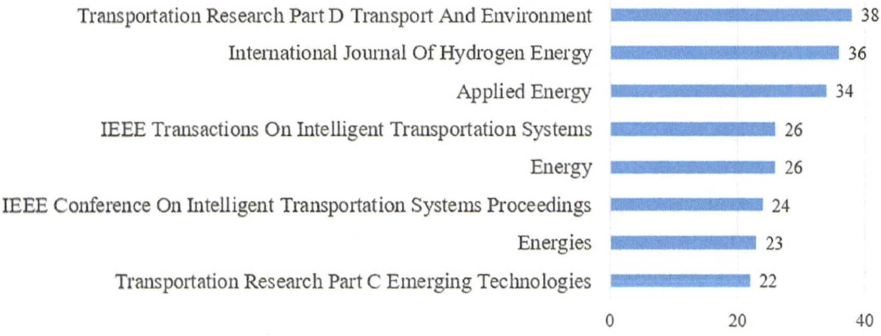

Fig. 4.2 Most prominent journals with publications on energy-related transport simulation (2010–2020)

nitrogen oxide (NO_X) emissions in tons (e.g., Auf der Landwehr et al. 2020). With a similar scope, Sayyadi and Awasthi (2018) built a simulation model in order to explore determinants for sustainable transportation planning, e.g., fuel consumption and greenhouse emissions. In this sense, they have tested several inputs within a System Dynamics (SD) framework, resulting in substantiated recommendations to decision makers about defining guidelines to reduce urban travel distances and trip rates.

Since 2007, electric vehicles (EVs) have gained an increasing interest (Juan et al. 2016). Wu and Zhang (2017) elaborate on this issue in great detail. In their article, the authors employ simulation to answer the question whether the development of EVs can reliably reduce air pollution and greenhouse emissions. The answers depend on the market energy mix. Here, simulation plays a critical role in predicting the future environmental effects caused by developing a country's EV industry. Two other examples are described in Moghaddam et al. (2017) and Wang et al. (2016). In these papers, the problem of identifying the optimal locations for EV charging points is addressed. In the former study, a network in Washington City was simulated with a population of up to 10,000 EVs, including various influencing parameters (e.g., arrival times, locations and number of charging points, energy consumption). In the latter research, simulation was used to generate and evaluate multiple public bus route networks with different battery sizes for individual buses.

Furthermore, cooperative strategies are simulated to assess the performance of different scenarios. This is further discussed and defined by Serrano-Hernandez et al. (2017). They distinguish two meta-approaches: (1) a consumer-centric approach, featuring sharing strategies such as carsharing and ridesharing, and (2) a company-centric solution based on horizontal cooperation.

Concerning ridesharing and carsharing, even though they are not closely related to logistics, they are cooperative passenger transportation strategies that are gaining momentum. Tikoudis et al. (2021) and Luna et al. (2020) provide a detailed elaboration on the respective effects and consequences in various contexts. In the former study, the authors estimate the proliferation effects of ridesharing services on transport-related CO_2 emissions in nearly 250 cities for the period 2015 to 2050. Their simulation models project a reduction of 6% in CO_2 emissions with ridesharing proliferation, while also outlining major differences depending on the individual city characteristics. In the latter study, the authors considered a SD simulation model to explain the impacts of electric carsharing on CO_2 reductions in the city of Fortaleza (Brazil). They concluded that carsharing schemes can play a vital role in reducing CO_2 emissions and improving urban mobility within the near future. In terms of horizontal cooperation, companies share their logistics resources with the purpose of obtaining financial and environmental benefits (e.g., by achieving economies of scale). Some examples are described in Serrano-Hernandez et al. (2018) as well as Fikar and Leithner (2021), where agent-based models are designed to track coalition forming and its influence on key performance indicators like driving distances, CO_2 emissions, and service quality. This particular case will also be addressed in the case study in Sect. 4.4.2.

It becomes obvious that simulation in transportation is gaining momentum in scientific literature. Commonalities across almost all simulation-based studies are the high degree of complexity regarding the underlying system as well as the conceptual characteristics and dimensions of the outcomes to be obtained. To this respect, energy-related indicators, mainly fuel (or energy) consumption and CO_2 emissions, are a must-have dimension for the dashboards of transportation managers. Finally, there is a rich range of simulation approaches, with agent-based modeling (ABM) coupled with Discrete Event Simulation (DES), and SD being the most prominent ones.

"Transportation systems and related policies are complex and cross-sectoral, covering different socioeconomic and management aspects" (Harrison et al. 2020, p. 239). To ensure a holistic, reliable, and integrated level of assessment and deduce feasible, system-specific recommendations, transportation-related problems require solution methodologies that offer different perspectives of transport planning while at the same time demonstrating the importance of systemic effects such as cause-and-effect relationships (Bierlaire 2015). Simulation constitutes a powerful and constructive tool to study the behavior of complex real-world transportations systems over time and evaluate energy-related system implications. In the context of transportation research, the term "simulation" is frequently used in a broad way. Hence, a precise definition of simulation technologies is required to distinguish this approach from a methodological point of view from other solution techniques such as static calculations, artificial neural networks, and fuzzy logic. The German Verein Deutscher Ingenieure (2014, p. 3) defines simulation as the "representation of a system with its dynamic processes in an experimentable model to reach findings which are transferable to reality." Simulation is characterized by experiments that provide information on one possible system or system variant. Therefore, it does not provide any best or optimal solution directly, but needs to be controlled manually or automatically to guide solution assessment over a series of experiments. Instead, simulation is particularly appropriate for performing "what-if" analyses on a given system (Rabe and Goldsman 2019).

Simulation methodologies describe the overall process of employing one or multiple simulation methods in terms of technology and procedures required to build a simulation model (a mathematical formulation that captures a system's simplified representation) and conduct simulation experiments on the system under study (Winsberg 2003). As such, a simulation methodology can be interpreted as a contextual framework that guides planning and handling of a simulation task or project. The choice and application of an appropriate methodology is inevitable to conduct reliable, efficient and valid simulation on energy-related questions in the transportation sector. In research and practice, various modeling and simulation methods featuring distinct benefits and limitations in specific application domains have evolved over the last decades (Diallo et al. 2015). With regard to the transportation sector, the most commonly employed methods include Monte Carlo Simulation (MCS), DES, and SD. While DES models a given business process as series of events with individual entities traversing these events, SD focuses on flows around networks rather than on the individual behavior of entities, continuously tracks system response according to a set of equations, and can mathematically be described as a system of differential equations (Morgan et al. 2017). Some problems even feature a set-up, where events occur at fixed times or where the exact time of occurrence is not relevant. In these cases, it may be appropriate to employ a time-slice simulation (TSS), which is explained in more detail in Chap. 1 of this book. Unlike dynamic methodologies such as DES and SD, static approaches such as MCS do not take into account system state changes that dynamically develop over time, but are a mere representation of a system's state at a given point in time. Hence, MCS is used to determine the behavior of a system by means of random samples and statistical evaluation (Mooney 1997). Moreover, ABM constitutes a popular approach to model a system under study. This modeling technique is generally coupled with DES or TSS methods and models a system as a network of autonomous agents following a set of predefined rules and conditions to draw conclusions from the individual agent behavior and interactions with the environment on the behavior of the entire system (Siebers et al. 2010). Finally, application specific techniques such as traffic simulation describe a group of different simulation approaches that have been specifically developed to solve traffic management problems. Thus, traffic simulation is not a simulation method per se, but rather a collection of domain-specific solutions. The system state, a set of variables that contain information on the evolution of the system over time, can both be based on discrete and continuous simulation methods. Depending on time-advancing-mechanism, system state and solution space, traffic simulation builds on various simulation methodologies such as cellular automata (discrete time, discrete state, discrete space) or numerical partial differential equation modeling (discrete time, continuous state, discrete space) (Barceló 2010).

4.3 Simulation Methodologies for Energy-Related Transportation Aspects

Table 4.1 provides an overview about the applications of different simulation methodologies when it comes to transportation management and planning. It is worthwhile to note that the simulation of energy implications in large and detailed transportation systems requires a considerable amount of computational resources, which can represent a limit to the dimension and the level of granularity of the simulation itself. However, new computational paradigms, as parallel and cloud computing, currently permit to even simulate enormous transportation and energy systems in a detailed and realistic manner (Lu and Zeng 2014).

Due to the fact that energy-related questions in the transportation sector are typically also characterized by a high degree of complexity, interdependency, and variability, both time-driven (TSS) and event-driven (DES) simulation approaches are

Table 4.1 Applications of simulation methods in the transportation sector

Methodology	Transportation domain	Examples
DES	Urban traffic management Concrete transport	Saltzman (1997) Khanh and Kim (2020)
DES with ABM	Railway transport Road transport City logistics	Böcker et al. (2001) Auf der Landwehr et al. (2020) Trott et al. (2021)
Continuous simulation	Traffic control City logistics	Boel and Mihaylova (2006) Simoni and Claudel (2018)
Hybrid (SD&DES)	Traffic control Road transport	Tako and Robinson (2012) Abzuaziz et al. (2015)
Hybrid (Others)	Urban traffic management	Zhang et al. (2014)
Static methods	Airport management Travel navigation and routing	Pitfield et al. (1998) Juan et al. (2011)

typically not capable to holistically evaluate energy requirements and implications across all system levels of traffic control and road-transport problems with large-scale networks in isolation. In this case, DES can be coupled with SD in a hybrid simulation to capture different abstraction levels and improve computational performance (Brailsford et al. 2014).

Most energy-related simulation projects in the transportation sector address the need to assess or optimize traffic related emission outputs such as carbon monoxide (CO), hydrocarbons (HC), CO_2, nitrous oxide (N_2O), ammonia (NH_3), NO_X, fine particulate matter ($PM_{2.5}$), and coarse particulate matter (PM_{10}). Other energy aspects include performance evaluation of energy sources such as biofuel, electricity, fossil, or hybrid systems (e.g., De Fillipo et al. 2014) as well as the consideration of energy policies and their impacts on macroeconomic drivers in different nations (e.g., Aslani et al. 2014). Since energy-related concerns generally encompass a macroscopic level of assessment and continuous system states (e.g., Abzuaziz et al. 2015), in recent years, hybrid methodologies have evolved as popular approach to tackle energy questions in the transportation sector (Brailsford et al. 2019).

4.4 Applications

To demonstrate the applicability of simulation methodologies for assessing energy-related aspects in the transportation domain, the following section elaborates on two exemplary use cases. These cases employ an ABM technique to assess emission implications resulting from different freight (Sect. 4.4.1) and grocery (Sect. 4.4.2) transportation strategies such as customer self-collection and horizontal collaboration, ultimately identifying conceptual characteristics as well as system designs that yield a high emission savings potential.

4.4.1 Last-Mile Parcel Deliveries in Hanover

The growing population, the increasing importance of the e-commerce business as well as the urbanization are putting a major strain on the infrastructure of cities and present logistics in urban areas with unprecedented challenges and requirements. Nowadays, parcel deliveries account for a major share of environment- and traffic-related issues in metropolitan areas (Van Duin et al. 2016). Courier, express, and parcel (CEP) deliveries are directly associated with the growing popularity of e-commerce and inherently responsible for the increase in urban road traffic and traffic-pertinent pollution (Pronello et al. 2017). Consequently, new ideas, concepts, and innovations are required to improve the operational efficiency of last-mile parcel deliveries and ensure more sustainable practices across the entire order-fulfillment-chain. Urban logistics concepts of the future need to be shaped in such a way that they compensate for both, the increasing demand for delivery convenience and velocity as well as the need to decrease environmental pollution and traffic emissions (Cullinane 2009). In this context, multiple last-mile parcel delivery strategies have evolved over the past decades, including dedicated parking concepts for carriers (e.g., Trott et al. 2021), consumer self-collection services (e.g., Yuen et al. 2018), crowdshipping approaches (e.g., Simoni et al. 2020) and collaborative white-label solutions (e.g., Pufahl et al. 2020). These concepts consist of a set of operational values for storage, transport, and handover routines such as central depots and local micro depots, delivery vans and cargos bikes, as well as attended and unattended home deliveries, which denote their individual implications (Boysen et al. 2021).

However, from an energy-related point of view, these implications are subject to a wide variety of partially interdependent influencing factors like depot locations, fleet compositions, routing procedures (Koç et al. 2016), and vehicle speeds (Demir et al. 2014), impeding the possibilities to quantify and analyze traffic emissions associated with last-mile parcel delivery concepts elaborately and correctly. In this regard, simulation modeling offers a feasible, rigorous and scalable opportunity to capture energy-specific ramifications, since it is capable of generating a multitude of virtual cases (Russo and Comi 2011), collect a vast amount of data required to compare and evaluate cross-parametric dependencies, sensitivities, and effects, and allows for studying phenomena in cases where it is intractable to conduct real-life studies due to prohibitive costs or conditions (Pidd 2004). Thereby, as an abstraction of reality, a simulation model must be constructed in a structured and formalized way to ensure that all relevant features and attributes of the real system are carried over to the virtual model during the abstraction process (Kotiadis and Robinson 2008). Moreover, accurate input data are required to precisely model process characteristics and operational peculiarities such as carrier distribution volumes or distances between locations (Balci 2012). Ultimately, the practical implications of the results of a simulation study are only as meaningful as the corresponding simulation model is an adequate reflection of the real system under investigation.

Using a case study for the city of Hanover in Germany, the quantification of energy-related traffic effects and emissions from different parcel delivery concepts is demonstrated. Taking into account the outlined necessity for structured and representative model building as well as data accuracy, in the following sections, the researched scenarios and the simulation modeling process as well as the model's parameters are outlined in detail. On the basis of these inputs, multiple simulation-based what-if scenarios are presented to compare the environmental consequences of different last-mile distribution strategies for CEP service providers.

4.4.1.1 Parcel Delivery Concepts

This simulation study is based on the daily distribution activities of a major LSP in Germany. The carrier's current operations model in the city of Hanover was assessed by means of interviews with drivers and managers, datasets on delivery destinations and travel times (e.g., tour schedules), as well as several field studies. As of April 2022, the total amount of parcels handled for the four investigated areas of

Hanover (Mitte, List, Groß-Buchholz, Oststadt) equals 8,600 per day. As conceptually depicted in Fig. 4.3, different CEP providers (C1–C3) ship parcels from central depots to regional depots in close proximity to the target area in Hanover, from where these are distributed on the last mile by light delivery vans. This model serves as benchmarking scenario for evaluating the implications of four alternative CEP distribution schemes in terms of traffic emissions. These alternative schemes encompass collaborative *white label* deliveries, last-mile distribution activities based on an extensive *micro hub* network, joint order fulfillment from a central *city hub*, and *self collection* from parcel shops.

Within the context of the *White-Label* concept, orders from several CEP service providers are bundled and carried out by a joint service in the city center (Fig. 4.4). The horizontal cooperation and the bundling of orders in joint regional depots may enable CEP providers to achieve greater economies of scale, which, in turn, is likely to exert a positive effect on traffic flow and emissions outputs. While regional depots are supplied individually by each carrier with heavy duty trucks, light commercial vehicle fleets are used jointly in order to utilize individual vehicles more effectively and minimize the individual distance driven per vehicle. Similar to traditional delivery approaches, the parcels are shipped to a priory specified location (home-delivery) and the parcel reception is unattended.

Concerning the *city hub* concept, a stationary, inner-city transshipment point is employed by several CEP LSPs for distribution activities on the last mile (Fig. 4.5). Accordingly, the model corresponds to a *white label* solution with individual, carrier-specific delivery routines. Customers are supplied by various CEP service providers via the stationary hub in the city center. The *city hub* serves as transshipment point, from where CEP service providers distribute parcels on the last mile via cargo bikes and light delivery vans. The hub features facilities and equipment required for vehicle parking, battery loading, and maintenance. End customers are supplied from the stationary hub by attended home deliveries.

The *micro hub* concept encompasses multiple container-sized hubs across the city area, which will be used as final point of fulfillment for distribution activities on the

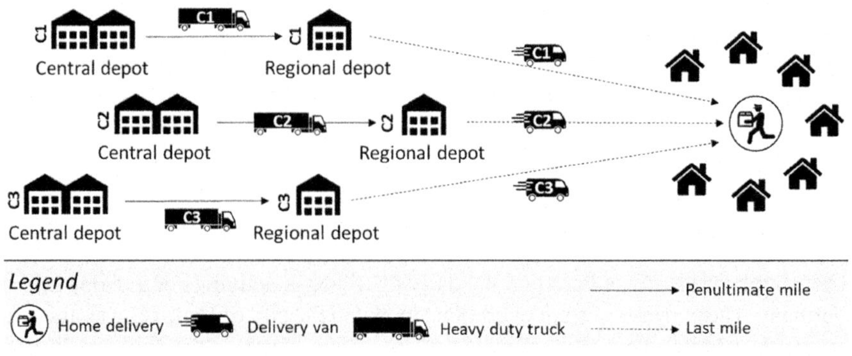

Fig. 4.3 Traditional parcel delivery concept

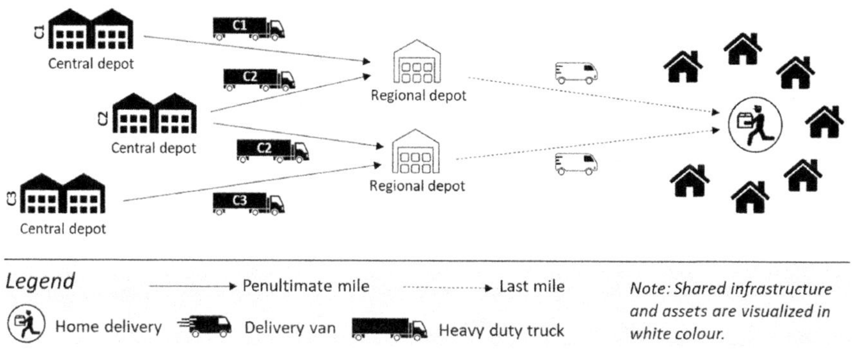

Fig. 4.4 White label parcel delivery concept

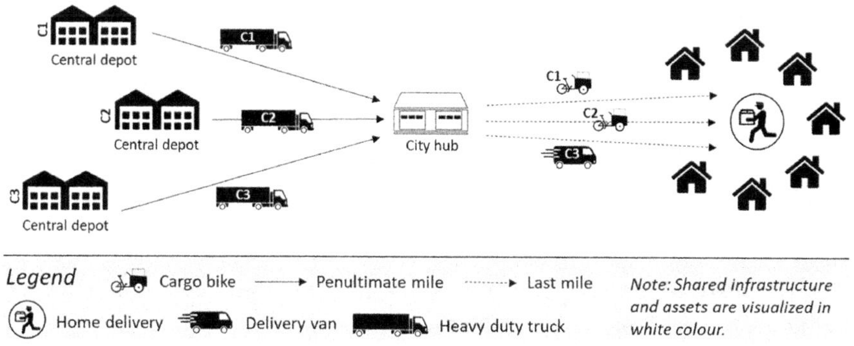

Fig. 4.5 City hub parcel delivery concept

last mile (Fig. 4.6). Parcels are delivered individually by each CEP service provider and *micro hubs* are supplied once a day. The hubs enable last-mile deliveries by cargo bikes within a restricted radius of 500 m. Shipping activities exceeding this range are performed with delivery vans. The aim of this concept is to reduce emissions accruing on the last mile, reduce road traffic, and improve the overall traffic flow by reducing second-row parking instances. Since micro hub and cargo bike capacities are restricted compared to regional depots, city hubs, and delivery vans, a sophisticated hub network is required to ensure full coverage of the city.

The *self collection* delivery process resembles the *traditional* delivery concept (Fig. 4.7). However, in this case, CEP service providers waive home deliveries and provide parcels via a dedicated reception infrastructure. Reception points can include parcel shops, public stations (e.g., gas station, train station), as well as locker facilities, and require customers to bridge the last mile themselves. It is assumed that a correspondingly high density of storage locations is available, so that customers do not necessarily require motorized vehicles to collect a parcel, thus relieving traffic and reducing emissions.

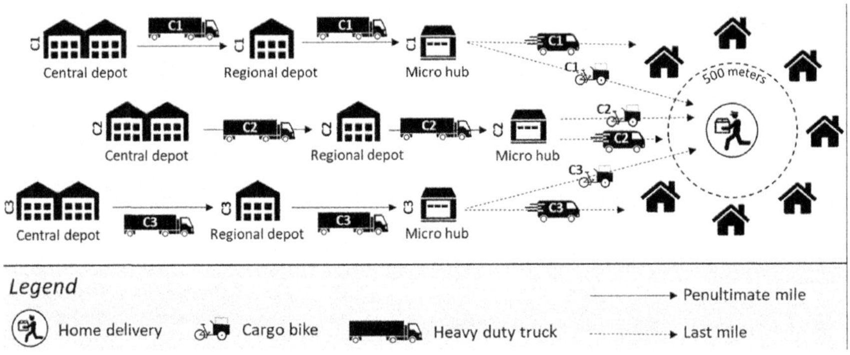

Fig. 4.6 Micro hub parcel delivery concept

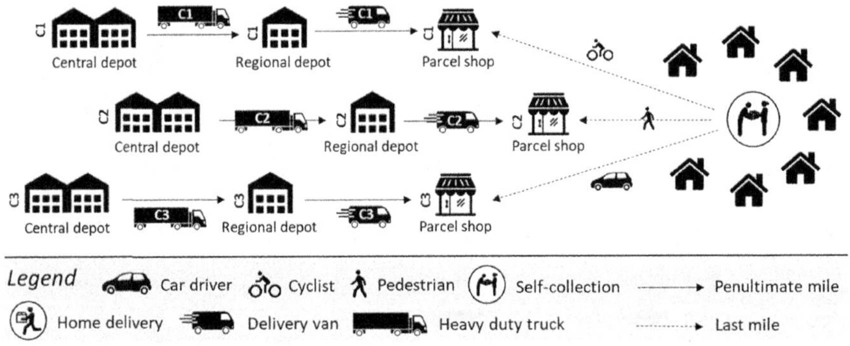

Fig. 4.7 Self collection parcel delivery concept

4.4.1.2 Modeling and Simulation Approach

The scope of this simulation study is restricted to four representative urban districts in the city of Hanover, Germany, featuring a total population of 9,400 inhabitants. For each delivery concept, different scenarios regarding infrastructural (e.g., hub capacities) and operational (e.g., shipping radii) peculiarities are investigated. Since this study focuses on the examination of last-mile distribution and to improve the comparability of the different concepts, the simulation model solely covers shipping activities from regional depots to final order recipients, consequently excluding supply activities by heavy duty trucks. Concerning the *city hub* concept, heavy duty truck mileages are merely tracked within an area that equals the outer radius of the target area and the regional depots of the *traditional* delivery concept. Trip distances are calculated with a bidirectional A* point-to-point algorithm (Nannicini et al. 2012) based on an OpenStreetMap network and validated with geographic data for the simulated city districts. To account for the fact that CEP delivery destinations and routes change day by day, the individual delivery frequency of a parcel recipient is modeled as a black box with fixed shares, whereby the individual recipients are stochastically altered with each simulation run (Monte Carlo approach). Similarly, several probabilistic system parameters such as vehicle speeds are varied based on stochastic distributions and computed as the average over a total of 180,000 simulation runs. Depot locations and parcel collection stations have been aligned with the existing infrastructure of five major CEP providers in Hanover, while micro hub and city hub locations were set based on the objective to minimize the distance to all potential recipients. A synopsis on the model input parameters used for this simulation study is provided in Table 4.2.

To develop the simulation model, the multimethod software AnyLogic (Version 8.7.1) was used. The simulation methodology combines ABM properties with a discrete event paradigm, whereby the synchronous time-advancing mechanism is triggered by sequential behavioral state changes of agents and the resulting interactions in the specified agent networks (DES). Behavioral rules for individual agents are modeled by state charts, defining the logical system flows, interdependencies and interactions based on the modeled state. This procedure allows for effectively modeling and representing the autonomous and heterogeneous behaviors of individual system entities (e.g., consumers), while taking into account collective interdependencies, and emerging reciprocations (Gómez-Cruz et al. 2017). The conceptual logic of the simulation model builds entirely on ABM and models each entity of the system (e.g., delivery van, order recipient) as individual agent or group of agents.

Each virtual simulation run equals one day and the model scope is limited to a total of 10,525 potential recipients that have been distributed randomly across the area of investigation. Physical agents (e.g., delivery vans) are placed in a geospatial environment, where distance-based navigation and routing procedures are conducted in line with a cluster- and time-window-based k-Nearest-Neighbor algorithm (Dudani 1976). The general shipping process is initiated by recipient-specific parcel orders, which are pre-sorted in the responsible depot. Subsequently, tours are created and assigned to the respective delivery fleet. Finally, depending on the individual time slot of a set of parcels, delivery tours are planned and conducted. If a parcel cannot be delivered successfully, it will be returned to the distribution center at the end of the tour and delivered the next day. Moreover, the individual capacities of different vehicle types were stochastically varied to account for different parcel size compositions.

To calculate the energy distribution in terms of emission outputs, simulated distance metrics are tracked for all vehicles. The referenced commercial vehicles are a Mercedes Benz 310 cdi/4325 (Delivery van; 70 kW/95 PS; Diesel; Euro 6b; 2.21 tons tare weight) and a MAN TGS 41.330 with Krone Profi Liner SDP 27 eLB4-CS (Heavy duty truck; 264 kW/360 PS; Diesel; Euro 6; 15.9 (tractor) + 6.2 (trailer) tons tare weight). As outlined in Eq. 4.1, emissions caused by private and commercial traffic ($E_{i,j}$) are calculated by the number of vehicles in the investigated area of category j and technology k ($N_{j,k}$), the average annual distance driven per vehicle of category j and technology k in kilometers ($M_{j,k}$), and the technology-specific emission factor of pollutant i for vehicle category j ($EF_{i,j,k}$). Vehicle categories include passenger cars, light commercial vans, and heavy duty trucks, while technologies

Table 4.2 Model parameters

Category		Value	Unit	Type	Concept
Delivery van fleet		42	Vans	Fixed	All concepts
Delivery van capacity (mean/SD)		160 (20)	Parcels	Stochastic	All concepts
Heavy duty truck fleet		20	Trucks	Fixed	All concepts
Heavy duty truck capacity		600	Parcels	Fixed	All concepts
Shipping volume per day		17,525	Parcels	Fixed	All concepts
Vehicle speed inner city (mean/SD)		25 (5)	km/h	Stochastic	All concepts
Vehicle speed outer city (mean/SD)		80 (10)	km/h	Stochastic	All concepts
City hub capacity		18,000	Parcels	Fixed	City hub
Cargo bike speed (mean/SD)		15 (5)	km/h	Stochastic	Micro hub, city hub
Cargo bike capacity (mean/SD)		50 (10)	Parcels	Stochastic	Micro hub, city hub
City hub infrastructure		1	Hubs	Fixed	City hub
Delivery success rate (mean/SD)		90 (5)	Percentage	Stochastic	Traditional, white label, micro hub, city hub
Share of delivery time windows	12:00–15:00 15:00–18:00 08:00–12:00	40 30 30	Percentage	Fixed	Traditional, white label, city hub, micro hub
Share of cargo bike deliveries		0/25/50	Percentage	Variable	City hub
Micro hub capacity		400/800	Parcels	Variable	Micro hub
Micro hub infrastructure		472	Hubs	Fixed	Micro hub
Regional depot infrastructure		5	Depots	Fixed	Traditional, white label, micro hub, self collect
Share of micro hub deliveries		33/66/100	Percentage	Variable	Micro hub
Parcel station capacity		100/200/300	Parcels	Variable	Self collection
Reception radius per station		500	Meters	Fixed	Self collection
Share of self-collection fulfillment		100	Percentage	Fixed	Self collection
Share of car driver collection		20	Percentage	Fixed	Self collection
Share of pedestrian collection		40	Percentage	Fixed	Self collection
Share of cyclist collection		40	Percentage	Fixed	Self collection

range from Euro 1 to Euro 6. Regarding private traffic, the fleet has been specified by structural data for the given pilot districts (Landeshauptstadt Hannover 2020). NH_3, N_2O, NO_x, and CO are calculated by the given emission factors, whereas CO_2 emissions of vehicles k combusting fuel m are derived by Eq. 4.2, where $FC^{CALC}_{k,m}$ is the fuel consumption of the vehicles for the respective time period and $r_{H:C}$ as well as $r_{O:C}$ the ratios of hydrogen to carbon and oxygen to carbon in the fuel. Input values on emission factors, vehicle categories, technologies, and pollutants were compiled from the European Environment Agency (Ntziachristos and Samaras 2019).

$$EP_{i,j} = \sum_k (N_{j,k} \times M_{j,k} \times EF_{i,j,k}) \tag{4.1}$$

$$E^{CALC}_{CO_2,k,m} = 44.011 \times \frac{FC^{CALC}_{k,m}}{12.011 + 1.008 r_{H:C,m} + 16.000 r_{O:C,m}} \tag{4.2}$$

4.4.1.3 Results

The simulation results can be interpreted from two different points of view: (1) the accruing traffic volumes in terms of mileages and (2) the associated emission outputs. As outlined in Fig. 4.8, total mileages are lowest for the *micro hub* concept and highest for the *traditional* delivery scenario, which is congruent with existing studies for other areas (e.g., Bergmann et al. 2020; Pufahl et al. 2020). Moreover, distribution strategies based on *self collection* and a *white label* paradigm seem to be a viable solution to decrease traffic volumes across the last mile in densely populated urban areas. Since the *micro hub* concept additionally comprises deliveries by means of cargo bikes, it is particularly favorable when it comes to reduce traffic volumes and road congestion. The *city hub* concept, which also includes the use of cargo bikes, accounts for less mileages driven than *traditional deliveries*, but performs worse than the priory mentioned concepts. Interestingly, the implications for concept-specific emission outputs are highly contradictory to the concept-implicated traffic volumes in some cases. Table 4.3 provides a synopsis on the individual emissions induced by each concept, highlighting the lowest and highest values for each pollutant.

Similar to the case of traffic volumes, the *micro hub* solution also seems to be highly effective when it comes to reducing traffic-related emissions. Furthermore, the *City-Hub* concept features comparably low traffic emissions in terms of CO_2, N_2O, NH_3, and CO. In contrast, due to the relatively high share of motorized traffic, the *white label* and *self collection* approaches feature relatively high degrees of CO_2 and N_2O emissions. *Self collection* even entails the highest emission outputs for NH_3 and CO pollutants across all concepts, which is mainly due to the differences in type of fuel and engine efficiency for private vehicles compared to commercial vans. While it is rather unfavorable in terms of CO_2 emissions, the *white label* solution is favorable from an ecological viewpoint, implying significantly lower NH_3 and CO outputs.

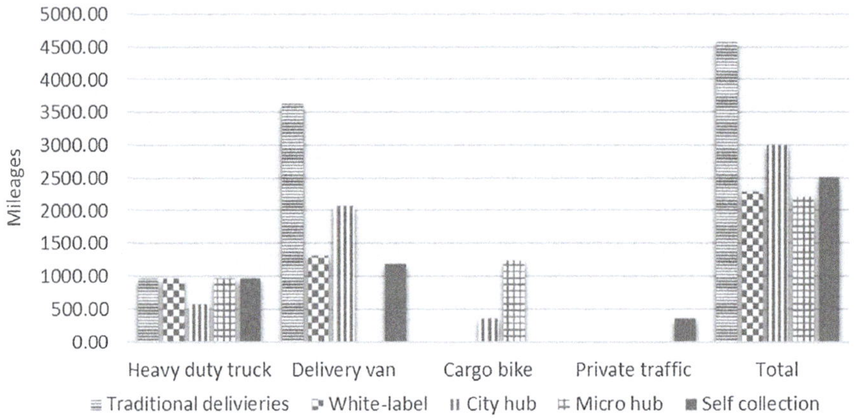

Fig. 4.8 Mileages per delivery concept (for the *city hub*, *micro hub*, and *self collection* with multiple instances in Table 4.4, mileages are shown for the instance with the lowest total amount)

Table 4.3 Total emissions per delivery concept (for the *city hub*, *micro hub*, and *self collection* with multiple instances in Table 4.4, mileages are shown for the instance with the lowest total amount)

Total emissions	Traditional	White label	City hub	Micro hub	Self collection
CO_2 emissions (in kg)	1839.87 ⇧	1256.32	921.45	860.50 ⇩	1294.90
NH_3 emissions (in g)	6.88	2.51 ⇩	4.79	2.81	9.31 ⇧
N_2O emissions (in g)	14.97 ⇧	5.77	0.29 ⇩	0.49	6.31
NO_x emissions (in kg)	3.52 ⇧	1.31	1.62	0.05 ⇩	1.26
CO emissions (in g)	271.67	99.03 ⇩	192.79	117.34	303.99 ⇧

Regarding the individual effects of different parameter variations across the *city hub*, *micro hub* and *self collection* concepts (each parameter configuration shown in Table 4.4 is referred to as individual concept instance), Table 4.4 indicates the potential to cut emissions by increasing the share of cargo-bike deliveries to decrease emissions from motorized vehicles (*City-Hub*) and employing a dense network of parcel stations with limited capacity (*Self-Collection*) to cut the share of motorized private traffic and raise the likeliness of collection activities by analog means of transport.

As for the *micro hub* concept, Fig. 4.9 illustrates that an extensive storage capacity of 800 and 100% hub-based deliveries in the target area are crucial to minimize both mileage and CO_2 emissions.

Table 4.4 Total emissions per delivery concept and concept instance

Main concept	Concept instance	CO_2 (kg)	NH_3 (g)	N_2O (kg)	NO_x (kg)	CO (g)
City hub	Cargo bike deliveries: 0 %	1130.74 ⇧	6.44 ⇧	285.12 ⇩	2.49 ⇧	257.80 ⇧
	Cargo bike deliveries: 25 %	1033.20	5.60	297.83 ⇧	2.02	224.85
	Cargo bike deliveries: 50 %	921.45 ⇩	4.79 ⇩	292.61	1.62 ⇩	192.79 ⇩
Micro hub	Micro hub capacity: 400; micro hub deliveries: 33 %	1143.63 ⇧	6.69 ⇧	265.77	2.67	267.22 ⇧
	Micro hub capacity: 400; micro hub deliveries: 66 %	1109.36	5.50	386.91	1.72	222.16
	Micro-Hub capacity: 400; micro-Hub deliveries: 100 %	878.45	2.87	501.93 ⇧	0.05 ⇩	119.79
	Micro- hub capacity: 800; micro hub deliveries: 33 %	1106.20	6.56	244.39 ⇩	2.66	262.12
	Micro hub capacity: 800; micro hub deliveries: 66 %	1098.92	5.47	380.95	1.72	220.74
	Micro hub capacity: 800; micro hub deliveries: 100 %	860.50 ⇩	2.81 ⇩	491.67	47.52 ⇧	117.34 ⇩
Self collection	Parcel station capacity: 100	1294.90 ⇩	9.31 ⇩	6.31 ⇩	1.26	303.99 ⇩
	Parcel station capacity: 200	1338.37	16.15	6.94	1.23 ⇩	510.40
	Parcel station capacity: 300	1404.38 ⇧	23.15 ⇧	7.92 ⇧	1.28 ⇧	723.48 ⇧

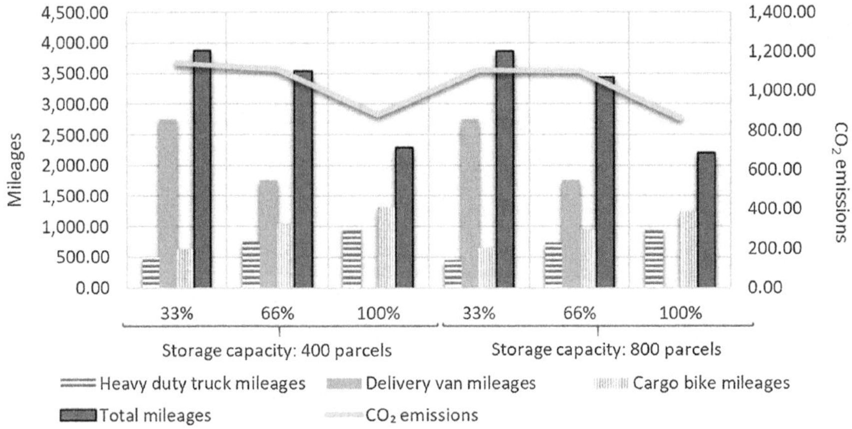

Fig. 4.9 CO_2 emissions (in kg) and mileages (in km) of micro hub concept instance

4.4.1.4 Discussion of Results

This use case compares five last-mile delivery concepts for CEP delivery services in terms of emission implications. To ensure a comprehensive and contrastable level of assessment, an ABS model has been developed with a discrete time-advancing mechanism for the city of Hanover (Germany), which is capable of reliably mimicking the given transportation system and its traffic ramifications. Based on this simulation model, the effects with regard to CO_2, NH_3, N_2O, NO_x, and CO pollutants have been evaluated for *traditional* home deliveries and compared them to *white label*, *city hub*, *micro hub* and *self collection* fulfillment concepts. For each concept, different scenarios in terms of fulfillment peculiarities (e.g., hub capacities) were tested to increase the resilience of the simulation's outcomes. Finally, the results indicate that *white label* operations seem to be particularly favorable to reduce emissions induced by CEP traffic, while *micro hub* and even *self collection* approaches lead to increased emission outputs compared to *traditional* CEP deliveries.

4.4.2 *E-grocery in Pamplona with Cooperation Strategies*

During the last decade, consumers' shopping habits have drastically changed, not only because of the massive incorporation of new technologies into our lives, but also because of a greater awareness of environmental and social sustainability, growing urbanization, and the increasing notion of time pressure. Actually, according to Eurostat (2021), 72% of EU citizens have bought or ordered goods or services online during 2019—prior to the COVID-19 pandemic—whereas that share was just 62% in 2015. In countries such Czech Republic, Romania, and Croatia, these increases reached around 20 percentual points in the same period. This is even more intense as a consequence of the pandemic, where online shopping surged, and new item categories gained importance such us food (for cooking at home) or personal care (Accenture 2020). The outcome is a challenging scenario, mainly driven by the increase in demand for e-groceries (i.e., the online purchase of groceries, including fresh products) because of an exceptional development of the e-commerce sector. Due to this paradigm shift in consumption, companies have adopted proactive sustainable strategies and developed sustainable supply chain management practices to respond to the evolving consumer preferences such as automatic parcel locker networks and electric vehicles, among others. However, and despite the complexity that is entailed with the growing demand, existing literature does not holistically demonstrate the challenges of the field, especially those related to logistics and fulfillment processes. Further, notwithstanding the promising development of the e-grocery business, the lack of interest in developing cost-effective operations is also evident in multiple cases, since there are only a few e-grocers that have been able to establish or expand profitable operations (Olsson et al. 2019). Thus, the challenges in e-grocery logistics range from a wide variety of food-safety-related issues to operational peculiarities like storage temperatures, including perishability over time (Fredriksson and Liljestrand 2015). In addition, environmentally responsible customer profiles must also be considered to assess consumption patterns based on preferences and local demand. Considering these consumer requirements, it is more than likely that consumers' requirements differ from the seller's desires. While consumers usually prefer products from an origin in close proximity with long expiry dates, sellers would economically benefit from handling routines that prioritize items with shorter shelf lives in order to reduce food waste (Fikar 2018).

In fact, many researchers recognize the strategic importance of sustainability in the management of supply chains as a hot topic in scientific literature and it is widely accepted that sustainability cannot be achieved by companies in isolation (Reyes-Rubiano et al. 2021). Accordingly, integration and involvement are required. Reinforcing the same idea, Soosay and Hyland (2015) plead that supply chain members operate in more dynamic environments, characterized by globalization, rapidly evolving technologies, and increased customer responsiveness. Therefore, more integrative and cooperative efforts are required to embrace the full potential of the supply chain and its characteristics. Likewise, horizontal cooperation may be paramount when opting to meet the requirements of customers and suppliers in an efficient and sustainable way, for example, by improving efficiency in logistics (Serrano-Hernandez et al. 2017). Therefore, the partnering sellers aim at increasing productivity through close cooperation, e.g., by optimizing vehicle capacity utilization, reducing empty mileage and cutting costs of auxiliary activities to increase the competitiveness of their logistics networks. In this regard, Cruijssen et al. (2007) have enumerated the potential benefits of cooperation as follows: (i) reduction of costs of transportation; (ii) improvement of service quality by reducing operation times and lost goods; (iii) diminution of environmental and social impacts; (iv) mitigation of risks; and (v) enhancement of market share. Consequently, extrapolating the previous benefits, horizontal cooperation might be particularly interesting for e-grocery, where a wide range of customers are widespread in big cities or in rural areas, generating long empty backhauls upon completion of the delivery activities. Here, the load factors can be improved by means of cooperation (i.e., supermarkets share their logistics operations) to reduce empty backhauls.

Therefore, this case study uses the e-groceries field in a medium-sized Spanish city within a cooperative supermarket setting to assess delivery performance in both economic and environmental sides.

4.4.2.1 E-groceries Market in Pamplona City

The interest in analyzing the e-grocery demand in a medium-sized city like Pamplona is two-fold. Firstly, the e-grocery penetration is lower, and customers' characteristics heavily differ from those in large cities, which are better covered by the literature (Mkansi et al. 2019). Secondly, the transportation infrastructure is usually poorer than in large cities, which makes transportation activities generally less efficient (Alvarez et al. 2018) and increases the importance for optimization. Therefore, the geographical scope of this experiment focuses on the area of Pamplona in Northern Spain, which includes a population of about 250,000 inhabitants. Figure 4.10 shows demand and supply points of the city, with pink dots representing demand locations and black triangles typifying the selected supermarkets for the simulation experiments.

A survey was conducted in the area with the purpose of gathering e-grocery demand information in the month March 2020, which was after the first Spanish

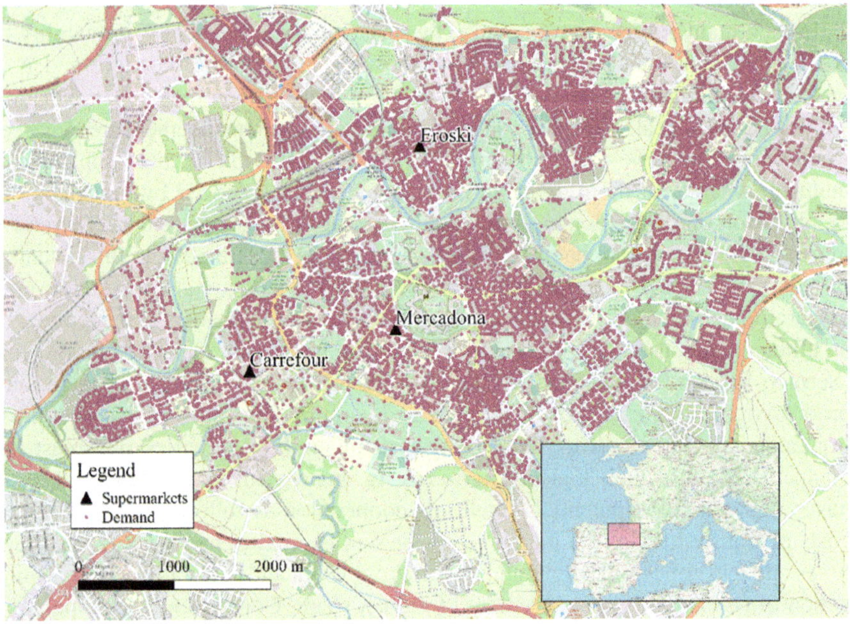

Fig. 4.10 Geographical scope of the case study indicating demand points and supermarkets

Table 4.5 Sample demographic counts

Age	Gender		Total
	Men	Women	
18–25	23	32	55
26–35	11	14	25
36–45	14	22	36
46–55	29	20	49
56–65	10	6	16
>65	0	1	1
Total	87	95	182

lockdown. The questionnaire comprised three question blocks: The introductory section aimed at collecting socio-economic information such as age, gender, and economic status. It is particularly meaningful and interesting as it introduces the topic and sets the tone for the remaining study. Therefore, the main objective of this section was to clarify general terms and concepts such as e-grocery demand. The second section was intended to gather operational e-grocery information. It contained questions related to supermarket preference, type of product, frequency of online grocery shopping, and the respective expenses. Finally, the third section is focused on the logistics part of the e-grocery service. Therefore, the questions here referred to the time dimension of the delivery service, namely the preferred day of the week for shopping, as well as the preferred time window for the delivery. The selection procedure was based on simple random sampling using email. For this purpose, different mail distribution lists, e.g., from the Council and the Public University of Navarra, were used to reach out to potential participants. All in all, 182 completed surveys have been collected. Main demographic figures are shown in Table 4.5. According to the latest available data on the distribution of the population in Pamplona by gender and age groups, the given sample slightly overrepresents the young and underrepresents the old but fits well to the gender distribution. Overall, the sample seems to be representative for the proposed case study.

From the analysis of the survey, the main points related to e-grocery demand patterns in the city of Pamplona can be drawn. First, there are three main supermarkets for ordering online. Thus, these three supermarkets are used for the simulation model (Table 4.6). Second, most of the participants usually do not order groceries online. However, about 25% of the consumers in the sample order e-groceries at least once a month. Third, deliveries are usually preferred on weekdays between 7 and 10 pm. The detailed delivery preferences including order frequencies are provided in Table 4.7.

With the previously obtained information, the expected demand, measured as the number of orders, can be estimated for the considered supermarkets. These estimations, as shown in Fig. 4.11, constitute the main input parameters for the simulation model.

Table 4.6 Consumers' supermarket preference

Supermarket	Preference (%)
Eroski	17.60
Mercadona	9.10
Carrefour	7.50

Table 4.7 Consumers' e-grocery ordering frequency

E-grocery ordering frequency (%)	
Once a week	4.87
Once every two weeks	8.70
Once a month	12.5
Once every two months	4.18
Once every three months	4.52
Never	65.2

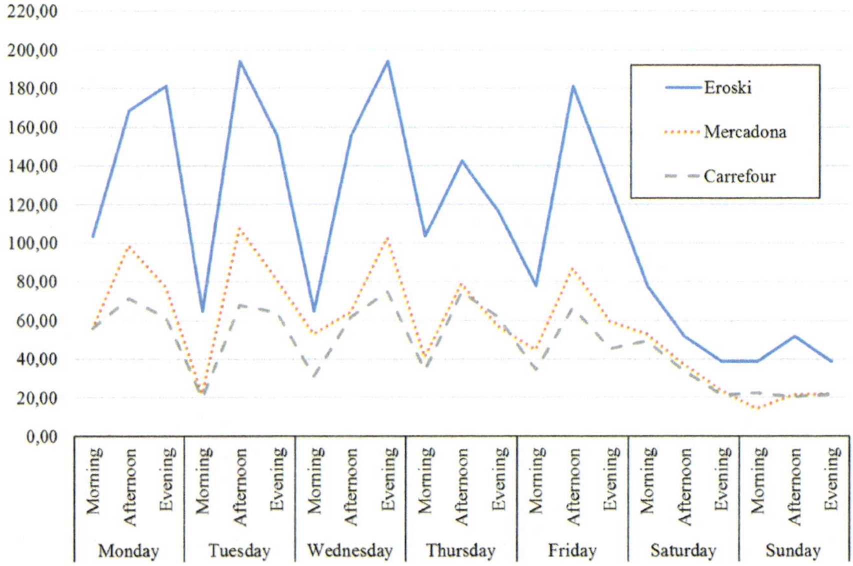

Fig. 4.11 Expected order demand per day, time window, and supermarket

4.4.2.2 Modeling and Simulation Approach

For analyzing the impact of horizontal cooperation on the urban e-grocery distribution in Pamplona, an agent-based model has been developed. As described in forthcoming subsections, the general idea behind the simulation model is that customers place orders to their preferred supermarket and indicate the preferred time-window for accepting a delivery. Afterwards, the stores have to fulfill the orders, depending on the run configuration, either by means of cooperative policies or individually.

A two-dimensional indicator has been applied to evaluate the impact of horizontal cooperation. The first dimension is an economic indicator measured as the total distance driven by vehicles. Secondly, an environmental indicator considers the resulting CO_2 emissions. With this respect, the methodology proposed by EcoTransIT (2019) for a small EURO VI truck with a 3,500 kg payload capacity was followed. Accordingly, energy consumption in MJ/km was estimated to be 4.7–5.1 for empty and full load operations in a standard scenario. Assuming that the vehicles in the experiments are fueled with Diesel, which features 43.1 MJ per kg and a density of 0.85 kg per liter of fuel, this leads to a total of 0.093–0.101 L of diesel consumption per kilometer. Moreover, utilizing the average conversion factor of 2.686 kg of CO_2 per liter of diesel, a final range of 0.249–0.270 kg of CO_2 emissions per kilometer is obtained.

The simulation model is built based on customer and supermarket agents. *Customer agents* are represented by the cadastral information in the area using a geographical information system (GIS). They are represented by the pink dots shown in Fig. 4.10, which outlines the location for each building of the 12,000 constructions in the metropolitan area of Pamplona. Knowing the population of the city and the size of each household (with an average of 2.5), the authors assume that each building lodges eight households. This simplification may result in weaker results for suburbs where the buildings account for a lower number of households. Therefore, in the simulation model, each demand point is replicated eight times. Parameters and variables associated with each of the roughly 96,000 demand points in the simulation model are related to the nature of the buyer, i.e., whether it is an e-grocery buyer, and, if so, his or her preferred supermarket, preferred time window and day of the week for e-grocery reception, as well as the lead time from the beginning of the selected time window and the moment at which the products are handed over to the consumer. Additionally, it is assumed that each customer has a service time of three minutes, whereby this time is considered as the temporal interval of making the physical delivery between the last mile distribution vehicle and the customer's home. For simplification purposes, that service time is fixed and the no-show share is set at 0%.

Supermarket agents typify the top three e-grocery supermarkets in Navarra, the Spanish region where Pamplona is located in (Eroski, Mercadona, and Carrefour). They are popular supermarket chains in Spain and offer a wide range of online groceries, including fresh vegetables and fruits. The locations of these supermarkets are highlighted in Fig. 4.10 as black triangles. The parameters and variables associated with each supermarket comprise the list of customers to serve each day and time window, driving distances, and the available fleet. The latter is a critical part in the logistics performance of the company. Hence, it is assumed that each supermarket owns a homogeneous fleet with a capacity of 20 orders. Likewise, the required size of the fleet has been determined with the expected weekly demand per time window, which is shown in Fig. 4.11. Considering all orders submitted to the supermarkets per day, average values have been obtained of 100.36, 51.89, and 42.77 orders for

Eroski, Mercadona, and Carrefour supermarkets, respectively. Hence, knowing the aforementioned demand values, the fleet size is set as four vehicles for Eroski and two vehicles for Mercadona and Carrefour, with the purpose of having a fleet size that is correlated with the number of orders in each supermarket.

If cooperation is not enabled, each supermarket will serve its customers in an independent manner. Consequently, each supermarket has to solve as many Vehicle Routing Problems (VRP) as time window slots it offers to design the orders distribution plans. Therefore, we have implemented a heuristic algorithm to solve each VRP, which is based on a biased randomization solution procedure of Clarke and Wright's Savings algorithm (Clarke and Wright 1964). A similar implementation considering sustainability dimensions and multicriteria analysis can be found in Abdullahi et al. (2021).

In the cooperative settings, all supermarkets serve all customers conjointly. The three supermarket chains form a coalition, which sets a delivery problem for the demanded orders. This problem must be solved considering several Multi Depot Vehicle Routing Problems (MDVRP) according to the time window slots that are given. Consequently, a heuristic MDVRP has been implemented following the recommendations described by Juan et al. (2015). The solution procedure starts by allocating the supermarkets to each customer based on time distances. Then, each customer is randomly assigned to a supermarket using a biased randomization procedure: closer supermarkets to the customers have greater probabilities to be chosen. Once all customers are assigned, the same biased-randomization procedure previously described in the VRP is applied to obtain a complete solution. Finally, this solution is saved, while a specific proportion of customers (65% in our experiments) remains unassigned and is subsequently reassigned by re-employing the biased-randomized assignment procedure. Then, the MDVRPs are solved again. This process is repeated 200 times, and the best solution so far is reported.

The dynamics of the simulation experiments are as follows. All parameters related to customer and supermarket agents are determined for each simulation replication. According to the input data, the customers place their e-grocery orders to their preferred supermarket to be served during a specific time window on a given weekday. Then, the two cooperation settings are tested. Thus, the simulation model starts on Mondays with the non-cooperation strategy. Orders are delivered following a sequential policy according to time windows. All the supermarkets start their deliveries at 7 am using the solution reported by the VRP algorithm. This is repeated for the rest of the week. Once the non-cooperative scenario is solved, the key performance indicators are returned, and the cooperative protocol is evaluated following the procedure previously described. The parameters set at the beginning of the replication are maintained for these settings. In total, we ran 100 simulation replications.

4.4.2.3 Results

The simulation model and the algorithms were implemented in AnyLogic 8.7.7 and run on a standard desktop with an Intel® Core™ i5-7400 CPU @3.00 GHz and 16 GB RAM.

First, a significant reduction in both, distances driven and CO_2 emissions can be observed when horizontal cooperation is employed. In particular, a 39.8% reduction in kilometers driven and 40.51% in CO_2 emissions can be expected. Table 4.8, based on 100 simulation runs, summarizes the main results on a weekly basis for the three supermarkets considered. Additionally, Fig. 4.12 outlines the boxplots for distances and Fig. 4.13 the CO_2 emissions for these simulation runs.

Additionally, a per-supermarket analysis has been performed. In this sense, Table 4.9 shows the impact on mileages and emissions of each supermarket depending on the cooperation strategy. This concept is further illustrated in Fig. 4.14. The analysis outlines significant differences depending on the size of a supermarket, highlighting in particular noteworthy savings for outlets with high demand shares.

Table 4.8 Impact of the cooperation in e-grocery delivery based on 100 simulation runs

Cooperation	Distances (km)	CO_2 (kg)
No	551.96	146.43
Yes	332.30	87.12
% Change	−39.80%	−40.51%

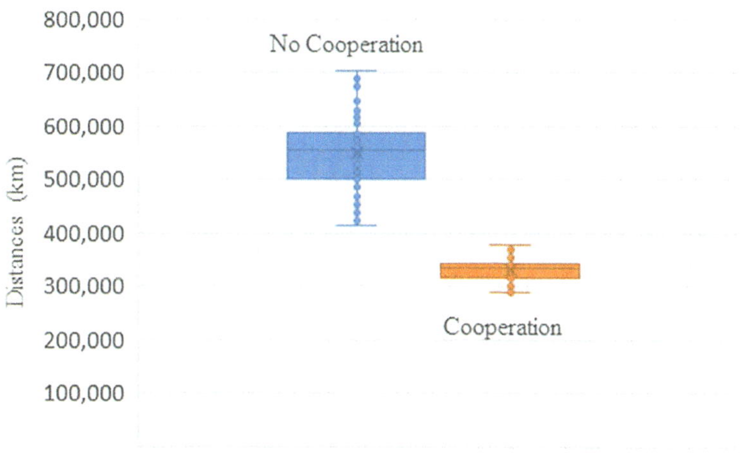

Fig. 4.12 Boxplots for distances driven (km) in cooperation and non-cooperation settings based on 100 simulation runs

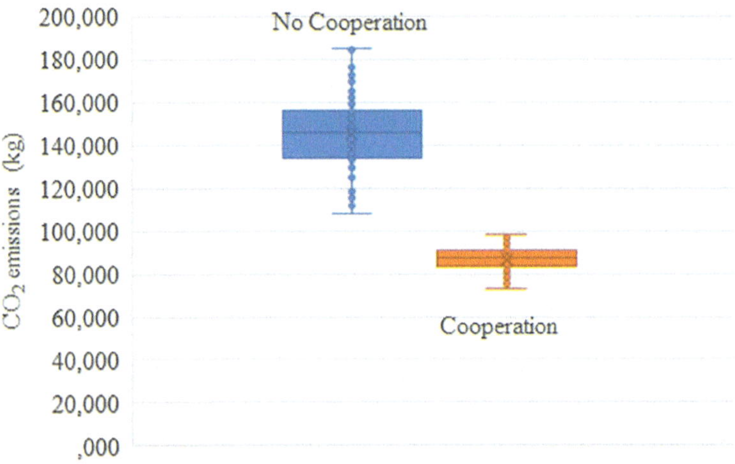

Fig. 4.13 Boxplots for CO_2 emissions (kg) in cooperation and non-cooperation settings based on 100 simulation runs

Table 4.9 Impact of the cooperation in distances driven (km) and CO_2 emissions (kg) per supermarket based on 100 simulation runs

Distances (km) and CO_2 (kg)	Cooperation				% Change	
	Yes (km)	Yes (CO_2)	No (km)	No (CO_2)	Km %	CO_2 %
Eroski	283.66	77.25	159.37	42.28	−43.82	−45.27
Mercadona	145.85	38.69	92.66	24.58	−36.47	−36.47
Carrefour	122.45	30.48	80.17	20.26	−34.53	−33.55
Total	551.96	146.43	332.20	87.12	−39.81	−40.50

4.4.2.4 Discussion of Results

This work presents the use of horizontal cooperation to gain competitiveness in the e-grocery delivery sector. For testing the convenience of using horizontal cooperation, an agent-based model for the city of Pamplona (Spain) has been developed. The evaluation focus was on the effects on economic measures and environmental implications of the logistics operations for different scenarios, which are based on the distribution of online demand orders for supermarkets. Two degrees of horizontal cooperation for performing the deliveries were tested, while distribution plans were determined by the implementation of a biased randomization algorithm. As a result, the simulation results indicate that the use of horizontal cooperation clearly improves the economic and environmental performance of e-grocery distribution activities.

This would have important considerations for the current supermarket business models, even if they are not out of step with the evolution of society as a whole. However, the exposed methodology has to be rolled out to other cities and businesses in order to generalize the conclusions found.

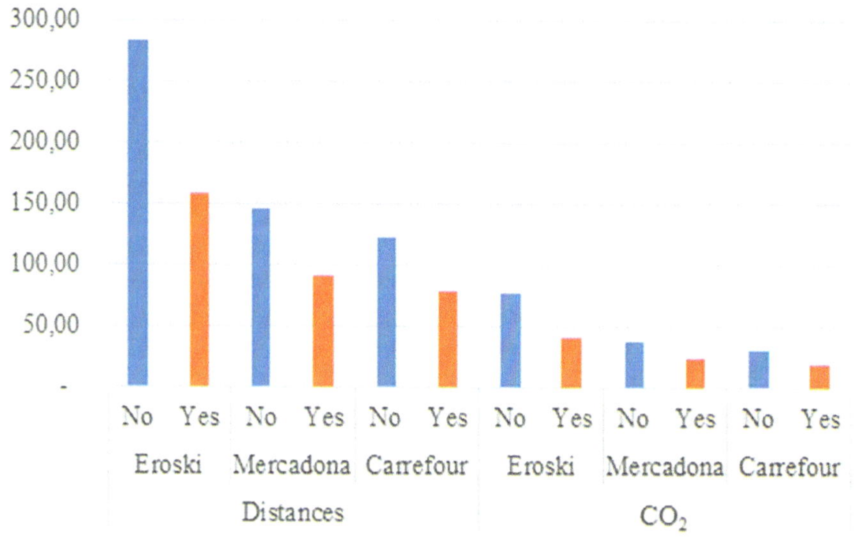

Fig. 4.14 Comparison of cooperation and non-cooperation settings for each supermarket in distances and CO_2 emissions

4.5 Conclusions and Outlook

This chapter elaborates on the use of computer simulation to study energy-related effects in the transportation sector. Freight transportation is a fundamental business determinant that typically features multiple contextual and conceptual trade-offs as well as interdependencies. To study transportation systems and their immanent energy implications, scholars require comprehensive, adaptive, and dynamic methodologies that are capable of capturing these relationships in a holistic manner. In this respect, computer simulation offers a valuable and distinctive tool for analysis, evaluation, and system design. In order to position the simulation methodology as an eligible assessment instrument for energy aspects in transportation systems, this chapter has first outlined the current status quo of energy-related simulation research, proving its growing popularity and usefulness in scientific research. Second, the characteristics of prevalent simulation methodologies in transportation science are synopsized to provide a reference framework for future simulation-based investigations in different related industry sectors. To demonstrate the applicability of dynamic computer simulation for assessing specific energy aspects in transportations such as traffic emissions, two dissenting use cases are outlined. The first case evaluates the energy efficiency of different fulfillment strategies in the courier, express, and parcel sector based on an operational setup in the city of Hanover (Germany). The second case elaborates on the economic and environmental effects of horizontal collaborations in the e-grocery sector, based on the operations of three major retailers in the city of Pamplona (Spain). Based on the multi-method software AnyLogic, these use cases illustrate both the innate capabilities of the simulation methodology for assessing energy-related aspects in the transportation sector as well as the multifariousness as to which simulation methods can be employed within this domain. Other tools commonly employed to assess similar transportation-related energy aspects are MATSim, PTV Vissim, SUMO and Simio.

The choice of modeling and simulation approaches as well as the resolution of investigated energy data mainly depends on the system and problem domain that is to be investigated. Transport-related energy concerns that directly relate to a multiplicity of individual behaviors and service-networks (e.g., influence of new public transport services on local traffic), can best be researched via an ABM paradigm that is coupled with a DES or TSS mechanism. If the investigation is to be conducted over a long period of time (e.g., impact of strategic transport policies at urban, regional and national levels), encompassing a high degree of time compression, SD can be a viable approach to investigate energy-related concerns. Finally, DES and TSS are particularly useful for examining the dynamics of transportation-related sub-systems that feature (e.g., operations of a single bus terminal).

Ultimately, recognizing simulation-based research as a powerful, valid, and effective method to investigate energy-related aspects in the transportation domain does not necessarily entail that simulation is always necessary for understanding every piece of every transport activity or system. However, it is an important step towards acknowledging this methodology as a viable, cross-disciplinary option for efficient and effective system design and analysis.

References

Abduaziz O, Cheng JK, Tahar RM, Varma R (2015) A hybrid simulation model for green logistics assessment in automotive industry. Procedia Eng 100:960–969

Abdullahi H, Reyes-Rubiano L, Ouelhadj D, Faulin J, Juan AA (2021) Modelling and multi-criteria analysis of the sustainability dimensions for the green vehicle routing problem. Eur J Oper Res 292(1):143–154

Accenture (2020) COVID-19: new habits are here to stay for retail consumers. https://www.accenture.com/hu-en/insights/retail/coronavirus-consumer-habits. Accessed 20 April 2022

Alvarez P, Serrano-Hernandez A, Faulin J, Juan AA (2018) Using modelling techniques to analyze urban freight distribution. A case study in Pamplona (Spain). Transp Res Procedia 33:37–74

Aslani A, Helo P, Naaranoja M (2014) Role of renewable energy policies in energy dependency in Finland: system dynamics approach. Appl Energy 113:758–765

Auf der Landwehr M, Trott M, von Viebahn C (2020) Simulation-based assessment of grocery shopping in urban areas. Simul Notes Europe 30(4):145–158

Balci O (2012) A life cycle for modeling and simulation. SIMULATION 88(7):870–883

Barceló J (2010) Models, traffic models, simulation, and traffic simulation. In: Barceló J (ed) Fundamentals of traffic simulation. Springer, New York, NY, pp 1–62

Bergmann FM, Wagner SM, Winkenbach M (2020) Integrating first-mile pickup and last-mile delivery on shared vehicle routes for efficient urban e-commerce distribution. Transp Res Part b: Methodol 131:26–62

Bierlaire M (2015) Simulation and optimization: a short review. Transp Res Part c: Emerg Technol 55(1):4–13

Böcker J, Lind J, Zirkler B (2001) Using a multi-agent approach to optimise the train coupling and sharing system. Eur J Oper Res 131(2):242–252
Boel R, Mihaylova L (2006) A compositional stochastic model for real time freeway traffic simulation. Transp Res Part b: Methodol 40(4):319–334
Boysen N, Fedtke S, Schwerdfeger S (2021) Last-mile delivery concepts: a survey from an operational research perspective. Or Spectrum 43(1):1–58
Brailsford S, Churilov L, Dangerfield B (2014) Discrete-event simulation and system dynamics for management decision making. John Wiley & Sons, Hoboken
Brailsford SC, Eldabi T, Kunc M, Mustafee N, Osorio AF (2019) Hybrid simulation modelling in operational research: a state-of-the-art review. Eur J Oper Res 278(3):721–737
Clarke G, Wright JW (1964) Scheduling of vehicles from a central depot to a number of delivery points. Oper Res 12(4):568–581
Corlu CG, de la Torre R, Serrano-Hernandez A, Juan AA, Faulin J (2020) Optimizing energy consumption in transportation: literature review, insights, and research opportunities. Energies 13(5):1115. https://doi.org/10.3390/en13051115
Crainic TG, Perboli G, Rosano M (2018) Simulation of intermodal freight transportation systems: a taxonomy. Eur J Oper Res 270(2):401–418
Cruijssen F, Dullaert W, Fleuren H (2007) Horizontal cooperation in transport and logistics: a literature review. Transp J 46(3):22–39
Cullinane S (2009) From bricks to clicks: the impact of online retailing on transport and the environment. Transp Rev 29(6):759–776
De Filippo G, Marano V, Sioshansi R (2014) Simulation of an electric transportation system at the Ohio State University. Appl Energy 113:1686–1691
Demir E, Bektaş T, Laporte G (2014) A review of recent research on green road freight transportation. Eur J Oper Res 237(3):775–793
Demir E, Huang Y, Scholts S, Van Woensel T (2015) A selected review on the negative externalities of the freight transportation: modeling and pricing. Transp Res Part e: Logistics Transp Rev 77:95–114
Diallo SY, Gore RJ, Padilla JJ, Lynch CJ (2015) An overview of modeling and simulation using content analysis. Scientometrics 103(3):977–1002
Dudani SA (1976) The distance-weighted k-nearest-neighbor rule. IEEE Trans Syst Man Cybern 6(4):325–327
EcoTransIT (2019) Ecological transport information tool for worldwide transports methodology and data update 2019. https://www.ecotransit.org/download/EcoTransIT_World_Methodology_Data_Update_2019.pdf. Accessed 20 April 2022
European Environment Agency (2021a) Transport data services overview. https://www.eea.europa.eu/themes/transport/dc. Accessed 20 April 2022
European Environment Agency (2021b) Greenhouse gas emissions from transport in Europe. https://www.eea.europa.eu/ims/greenhouse-gas-emissions-from-transport. Accessed 24 May 2022
Eurostat (2021) Online shopping ever more popular in 2020. https://ec.europa.eu/eurostat/web/products-eurostat-news/-/ddn-20210217-1. Accessed 20 April 2022
Faulin J, Lera F, Pintor JM, García J (2006) The outlook for renewable energy in Navarre: an economic profile. Energy Policy 34(15):2201–2216
Fikar C (2018) A decision support system to investigate food losses in e-grocery deliveries. Comput Ind Eng 117:282–290
Fikar C, Leithner M (2021) A decision support system to facilitate collaborative supply of food cooperatives. Prod Plann Control 32(14):1179–1190
Fredriksson A, Liljestrand K (2015) Capturing food logistics: a literature review and research agenda. Int J Log Res Appl 18(1):16–34
Gómez-Cruz NA, Loaiza Saa I, Ortega Hurtado FF (2017) Agent-based simulation in management and organizational studies: a survey. Eur J Manag Bus Econ 26(3):313–328
Harrison G, Bivona E, Rossetti R (2020) Editorial: special issue on simulation in transportation. J Simul 14(4):239–241

Janjevic M, Winkenbach M (2020) Characterizing urban last-mile distribution strategies in mature and emerging e-commerce markets. Transp Res Part a: Policy Practice 133:164–196
Juan AA, Faulín J, Jorba J, Riera D, Masip D, Barrios B (2011) On the use of Monte Carlo simulation, cache and splitting techniques to improve the Clarke and Wright savings heuristics. J Oper Res Soc 62(6):1085–1097
Juan AA, Pascual I, Guimarans D, Barrios B (2015) Combining biased randomization with iterated local search for solving the multidepot vehicle routing problem. Int Trans Oper Res 22(4):647–667
Juan AA, Mendez CA, Faulin J, De Armas J, Grasman SE (2016) Electric vehicles in logistics and transportation: a survey on emerging environmental, strategic, and operational challenges. Energies 9(2):1–21
Khanh HD, Kim SY (2020) Exploring productivity of concrete truck for multistory building projects using discrete event simulation. KSCE J Civ Eng 24(12):3531–3545
Koç Ç, Bektaş T, Jabali O, Laporte G (2016) The impact of depot location, fleet composition and routing on emissions in city logistics. Transp Res Part b: Methodol 84:81–102
Kotiadis K, Robinson S (2008) Conceptual modelling: knowledge acquisition and model abstraction. In: Mason SJ, Hill RR, Mönch L, Rose O, Jefferson T, Fowler JW (eds) Proceedings of the 2008 Winter Simulation Conference. IEEE, Piscataway, NJ, pp 951–958
Landeshauptstadt Hannover (2020) Strukturdaten der Stadtteile und Stadtbezirke 2019: Statistische Auswertungen in Form von Karten, Grafiken und Tabellen. https://www.hannover.de/content/download/800320/file/Strukturdaten2019.pdf. Accessed 20 April 2022
Lu G, Zeng WH (2014) Cloud computing survey. In: Yarlagadda P, Choi S-B (eds) Applied mechanics and materials. Trans Tech Publications, Zurich, ZH, pp 650–661
Luna TF, Uriona-Maldonado M, Silva ME, Vaz CR (2020) The influence of e-carsharing schemes on electric vehicle adoption and carbon emissions: An emerging economy study. Transp Res Part d: Transp Environ 79:1–14
Melkonyan A, Gruchmann T, Lohmar F, Kamath V, Spinler S (2020) Sustainability assessment of last-mile logistics and distribution strategies: the case of local food networks. Int J Prod Econ 228:107746. https://doi.org/10.1016/j.ijpe.2020.107746
Mkansi M, de Leeuw S, Amosun O (2019) Mobile application supported urban-township e-grocery distribution. Int J Phys Distrib Logist Manag 50(1):26–53
Moghaddam Z, Ahmad I, Habibi D, Phung QV (2017) Smart charging strategy for electric vehicle charging stations. IEEE Trans Transp Electrification 4(1):76–88
Mooney CZ (1997) Monte Carlo simulation. Sage Publications, Thousand Oaks
Morgan JS, Howick S, Belton V (2017) A toolkit of designs for mixing discrete event simulation and system dynamics. Eur J Oper Res 257(3):907–918
Nannicini G, Delling D, Schultes D, Liberti L (2012) Bidirectional A* search on time-dependent road networks. Networks 59(2):240–251
Ntziachristos L, Samaras Z (2019) EMEP/EEA air pollutant emission inventory guidebook 2019: 1.A.3.b.i-iv road transport 2019. European Environment Agency, Luxembourg
Olsson J, Hellström D, Pålsson H (2019) Framework of last mile logistics research: a systematic review of the literature. Sustainability 11(24):7131. https://doi.org/10.3390/su11247131
Pidd M (2004) Computer simulation in management science, 5th edn. John Wiley & Sons, Hoboken
Pitfield DE, Brooke AS, Jerrard EA (1998) A Monte-Carlo simulation of potentially conflicting ground movements at a new international airport. J Air Transp Manag 4(1):3–9
Pronello C, Camusso C, Valentina R (2017) Last mile freight distribution and transport operators' needs: which targets and challenges? Transp Res Procedia 25:888–899
Pufahl L, Ihde S, Glöckner M, Franczyk B, Paulus B, Weske M (2020) Countering congestion: a white-label platform for the last mile parcel delivery. In: Abramowicz W, Klein G (eds) International conference on business information systems. Springer, Cham, pp 210–223
Rabe M, Goldsman D (2019) Decision making using simulation methods in sustainable transportation. In: Faulin J, Grasman SE, Juan AA, Hirsch P (eds) Sustainable transportation and smart logistics. Elsevier, Amsterdam, pp 305–333

Reyes-Rubiano L, Serrano-Hernandez A, Montoya-Torres JR, Faulin J (2021) The sustainability dimensions in intelligent urban transportation: a paradigm for smart cities. Sustainability 13(19):1–20

Russo F, Comi A (2011) Model system for the ex-ante assessment of city logistics measures. Res Transp Econ 31:81–87

Saltzman RM (1997) An animated simulation model for analyzing on-street parking issues. SIMULATION 69(2):79–90

Sayyadi R, Awasthi A (2018) A simulation-based optimisation approach for identifying key determinants for sustainable transportation planning. Int J Syst Sci Oper Logistics 5(2):161–174

Serrano-Hernandez A, Juan AA, Faulin J, Perez-Bernabeu E (2017) Horizontal collaboration in freight transport: concepts, benefits, and environmental challenges. Stat Oper Res Trans 529(41):1–22

Serrano-Hernandez A, Faulin J, Hirsch P, Fikar C (2018) Agent-based simulation for horizontal cooperation in logistics and transportation: from the individual to the grand coalition. Simul Model Pract Theory 85:47–59

Siebers PO, Macal CM, Garnett J, Buxton D, Pidd M (2010) Discrete-event simulation is dead, long live agent-based simulation! J Simul 4(3):204–210

Simoni MD, Claudel CG (2018) A simulation framework for modeling urban freight operations impacts on traffic networks. Simul Model Pract Theory 86(1):36–54

Simoni MD, Marcucci E, Gatta V, Claudel CG (2020) Potential last-mile impacts of crowdshipping services: a simulation-based evaluation. Transportation 47(4):1933–1954

Soosay CA, Hyland P (2015) A decade of supply chain collaboration and directions for future research. Supply Chain Manage Int J 20(6):613–630

Tako AA, Robinson S (2012) The application of discrete event simulation and system dynamics in the logistics and supply chain context. Decis Support Syst 52(4):802–815

Tikoudis I, Martinez L, Farrow K, Bouyssou CG, Petrik O, Oueslati W (2021) Ridesharing services and urban transport CO_2 emissions: simulation-based evidence from 247 cities. Transp Res Part d: Transp Environ 97:1–23

Trott M, Baur NF, Auf der Landwehr M, Rieck J, von Viebahn C (2021) Evaluating the role of commercial parking bays for urban stakeholders on last-mile deliveries—a consideration of various sustainability aspects. J Clean Prod 312:1–15

Van Duin JHR, De Goffau W, Wiegmans B, Tavasszy LA, Saes M (2016) Improving home delivery efficiency by using principles of address intelligence for B2C deliveries. Transp Res Procedia 12:14–25

Verein Deutscher Ingenieure (2014) VDI Guideline 3633 Simulation of systems in materials handling, logistics and production—Part 1. Fundamentals. Beuth, Berlin

Wang X, Yuen C, Hassan NU, An N, Wu W (2016) Electric vehicle charging station placement for urban public bus systems. IEEE Trans Intell Transp Syst 18(1):128–139

Winsberg E (2003) Simulated experiments: methodology for a virtual world. Philos Sci 70(1):105–125

Wu Y, Zhang L (2017) Can the development of electric vehicles reduce the emission of air pollutants and greenhouse gases in developing countries? Transp Res Part d: Transp Environ 51:129–145

Yuen KF, Wang X, Ng LTW, Wong YD (2018) An investigation of customers' intention to use self-collection services for last-mile delivery. Transp Policy 66:1–8

Zhang B, Chan WKV, Ukkusuri SV (2014) On the modelling of transportation evacuation: an agent-based discrete-event hybrid-space approach. J Simul 8(4):259–270

Chapter 5
Retail

Kai Gutenschwager, Markus Rabe, Michael E. Kuhl, and Jorge Luis Chicaiza-Vaca

Abstract In the last decades, companies in the retail sector have faced growing customer demands for convenient delivery of products combining the alternatives of traditional in-store shopping and online shopping providing home deliveries or pickups at specific pickup locations, along with global aims to reduce energy consumption and CO_2 emissions worldwide with respect to the global climate change. Targets to reduce emissions are supported by a majority of the world's societies and companies. Especially retailers are characterized by very high transportation volumes with often very small transportation lot sizes. Here, distribution networks need to be designed that allow low energy consumption while still addressing customer demands. Simulation has been a core method to analyze such networks and underlying processes with respect to costs, but also enables detailed analyses of the energy consumption and CO_2 emissions of such systems. This chapter gives an overview of the scope and objectives for retail distribution systems and present challenges for respective simulation models, both addressing Discrete Event Simulation and System Dynamics. Furthermore, it presents three application studies and gives an outlook to future research and applications.

K. Gutenschwager (✉)
Institute of Information Engineering, Ostfalia University of Applied Sciences, Am Exer 2, 38302 Wolfenbüttel, Germany
e-mail: k.gutenschwager@ostfalia.de

M. Rabe
Fachgebiet IT in Produktion und Logistik, TU Dortmund University, Leonhard-Euler-Straße 5, 44227 Dortmund, Germany

M. E. Kuhl
Industrial and Systems Engineering Department, Rochester Institute of Technology, 81 Lomb Memorial Drive, Rochester, NY 14623, USA

J. L. Chicaiza-Vaca
Pichincha Chamber of Freight Transportation, Av. El Inca OE1-70 (N44-17), Quito, Ecuador

S. Wenzel et al. (eds.), *Energy-Related Material Flow Simulation in Production and Logistics*, https://doi.org/10.1007/978-3-031-34218-9_5

5.1 Introduction

Retail distribution systems are typically characterized by a relatively high energy consumption, as customers are to be served with several goods either directly (home deliveries) or by a network of stores or simple pickup locations, where customers shop or pick up online orders. For both alternatives, often small amounts of loading units need to be transported per order or per store. Transports could be carried out by a third party provider or an own fleet of vehicles. In this context, not only the size and composition of the fleet along with planning of tours is of interest, but also the design of the underlying network, consisting of suppliers, warehouses, hubs, stores, or pickup locations, as well as possibly the integration of third party logistic providers offering different modes of transport.

The remainder of this chapter is organized as follows: Sect. 5.2 provides an overview of the scope and objectives for the simulation of retail distribution systems. In Sect. 5.3, the specific challenges for setting up respective models are described in detail, also concerning the functionalities that the applied tools need to offer. In Sect. 5.4, three case studies are given. The chapter ends with a short conclusion and outlook for further developments (Sect. 5.5).

5.2 Scope and Objectives

A retail distribution system is a specific distribution system where the final customers are part of the respective system. Customers typically go to stores or other pickup locations or receive their goods at home. Often, retailers offer several possibilities to their customers, differentiating single and multi-channel—or even omni-channel—systems. The increase of the share of online sales both as a development in existing online channels and by the ongoing market entry of bricks-and-mortar retailers into e-commerce leads to non-linear fulfillment processes, as bricks-and-mortar retailing is increasingly overlapping with distance retail. Especially due to the COVID-19 pandemic, customer behavior has further shifted towards online shopping and home deliveries. According to data originating from Adobe, the USA have experienced a growth of 55% in online purchases, leading to a total of 1.7 Trillion US Dollars in the first two pandemic years (Koetsier 2022). Allianz has reported growth rates for 2020 and 2021 for a number of selected retail chains, where growth rates for online purchases often exceeded 100% in 2020, and still where partially in ranges near 100% in 2021 (Duthoit 2021). Hübner et al. (2016) give an extensive literature overview for retail distribution systems in omni-channel retailing. Here, the forward and backward distribution system are differentiated, where the forward distribution system covers the material flow from different sources, such as distribution centers, to customers or stores as possible points of reception. The backward distribution system deals with product returns from the customer to the retailer and possible return centers. This leads to complex distribution systems for both forward and backward processes that serve customers in stores and simultaneously offer home deliveries, as well as in-store return of online orders (Hübner et al. 2016). Von Viebahn et al. (2020) present a taxonomy for e-grocery fulfillment, differentiating 20 dimensions, including reattempted delivery, the return procedure, and detailed picking and routing dimensions. Using a cluster analysis, the authors identify six different archetypes of retailers, e.g., "regional champions" with dedicated distribution centers and attended reception or "collaboration experts" characterized by large delivery areas and collaborative fulfillment strategies with hybrid structures.

Moreover, for analyzing retail systems, customer movements to stores are usually not taken into account. Thus, the sinks for respective simulation models are usually the stores. However, for comparing the energy consumption of the overall system, the movements of customers to stores should also be considered, especially when comparing distribution systems where other supply chain designs include a direct delivery to the customer. Auf der Landwehr et al. (2021) present such a simulation study that compares six fulfillment strategies in the e-grocery sector. In their approach, customers either make single movements for each order, or set up a shopping list, which takes the utilization rate for grocery shopping into account and is fulfilled by a trip by car.

Choosing distribution channels is one of the main design decisions for retailers (Beck and Rygl 2015). In case of multi-channel systems, more than one supply chain needs to be planned in parallel. Here, different warehouse locations may be used for different channels, which increases the complexity respectively. In this context of supply chain design, typical planning tasks are (for more general supply chains, cf. Gutenschwager and Arnold 2020):

- the number of stores or pickup locations and their geographical locations,
- the number of (regional) warehouses, hubs, and possibly return centers and their geographical location,
- the assignment of articles and necessary storage resources (e.g., cooling systems) to given locations,
- the choice of mode of transport (including multi-modal transports) and service providers,
- the sourcing strategies (local sourcing vs. overseas transports or regional products), and
- the process of transport.

Typical key figures for supply chain decisions are transport costs, inventory costs, warehouse costs including costs for logistic processes, such as handling, repackaging or sealing. Most planning approaches for retail distribution systems address mainly these cost types. Energy costs may already be covered in these types of costs to some degree, and CO_2 emission strongly relates to other transport costs, mainly the driven distances and the respective fuel consumption, which will be further addressed in Sect. 5.4.1. Additionally, service-related key figures, such as different service levels towards the customers and response times (from order to delivery) come into play. These key figures are also offered by most of the respective simulation systems for supply chains (cf. Gutenschwager and Arnold 2020). In Hübner et al. (2016), these are also relevant key figures named by chief executives and supply chain experts in an empirical study. However, the authors do not include any aspects of energy consumption or green supply chains.

In this context, Susanty et al. (2016) present a general framework for green supply chain management practices. Green supply chain management is defined as the integration of environmental issues, covering all aspects from sourcing to distribution to the final customer, including reverse logistics (cp. Srivastava 2007; Sarkis et al. 2011). The energy consumption for warehouses, transport, and other processes (picking, storage, etc.) should be taken into account in more detail. Especially in the Fast Moving Consumer Goods (FMCG) sector—the typical product range for supermarkets—, there are typically zones in the warehouse with different temperature levels, e.g., for frozen goods, vegetables (near to zero) or dry storage. These sectors have different energy consumption levels, which need to be modeled in detail. This aspect might also have a strong impact on the overall distribution network design, e.g., by keeping frozen goods only at a centralized warehouse. As a general basis for the energy consumption, the CO_2 emission caused by the distribution system can be applied (Rabe et al. 2013).

5.3 Challenges for the Simulation Models

Modeling distribution networks is a highly complex task. Different simulation approaches can be found in the literature, mainly approaches in the field of System Dynamics (SD) and Discrete Event Simulation (DES). While SD is often used for rather general analyses, DES models are typically very detailed with models typically consisting of a large number of stores, warehouses, product groups, and suppliers. In the following two sections, the respective challenges will be discussed for the two approaches.

5.3.1 System Dynamics Models

System Dynamics (SD) is a methodology for modeling, simulation, analysis, and design of the dynamics of socio-economic systems and links qualitative and quantitative models. In SD, only a few basic modeling concepts are used: Causal feedback loops between system components can be modeled in respective causal loop diagrams. Such models are rather qualitative. In order to perform quantitative analyses, stock and flow diagrams can be used, which extend causal loop diagrams. Here, a flow leads to a—delayed—change of a stock (variable). These changes are typically modeled as equations. Compared to other simulation approaches, usually no complex logic models, such as algorithms to determine daily tours for a fleet of vehicles, are included. However, in case that the SD modeling environment is not capable of using certain algorithms, e.g., the calculation of expected stockouts, also an external program could be used to perform the required calculations (cf. Angerhofer and Anglidis 2000). Abbas and Bell (1994) conclude that SD is well suited for strategic problems and that it could provide a useful tool for supporting policy analysis and decision-making in the transport field, which is also highly relevant for retail systems. Shepherd (2014) identified several fields of application in the context of transportation for SD, reviewing more than 50 papers. Most of the problems dealt with in this overview rather have a strategic character and analyze long-term developments, like the uptake of alternate fuel vehicles, highway maintenance, and strategic policies at urban, regional, and national levels, and strategic supply chain management. Georgiadis et al. (2005) present a supply chain model based on SD for a fast food chain to investigate long-term capacity planning alternatives, as increasing the fleet size rather than leasing transport capacity. Das and Dutta (2013) present a simulation model to investigate the significance of various factors including product exchange, collection, and remanufacturing. The model also covers aspects of reverse logistics and refurbishing and gives insights to the well-known bullwhip-effect.

In Kazancoglu et al. (2021), a SD model has been developed to analyze and comprehend the green performance of reverse logistics activities by predicting the environmental impact in terms of CO_2 and other emissions. Figure 5.1 shows an extract of the causal loop diagram of the reverse logistics model presented.

Starting with the estimated total distance of routes—influenced by several factors including the country population—the total distance of routes is created on a monthly basis. Here, a forecast has been computed on historical data of more than 5,000 routes. The distances are then divided into one-way and round-trip distances. Furthermore, the number of routes and customer returns are estimated within the model. Once a one-way trip is planned, the return of that route is unplanned and creates emissions caused by non-value-adding activities. For round trips, different types of transportation tasks are differentiated, e.g., empty returns, pallet transportation, customer returns, or milk-run distribution, which all have different effects on CO_2 emissions, further differentiated into value-adding and non-value-adding activities (not shown in Fig. 5.1). The authors show that by decreasing the percentage of planned one-way-trip distances from 23 to 10% after five years, emissions caused by non-value-adding

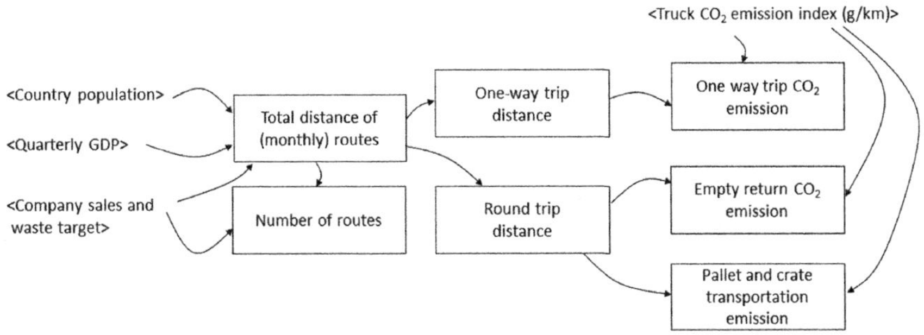

Fig. 5.1 Extract of a causal loop diagram of a reverse logistics model (Kazancoglu et al. 2021)

activities are reduced by 40%, while emissions caused by value-adding activities increase by 24%.

The main challenges when setting up SD models are the data preparation and defining valid flows. In the case study presented by Kazancoglu et al. (2021), an extensive network consisting of 149 distribution points is considered, from which the number of routes and distances to serve all customers has to be estimated for future scenarios in a valid way. Considering the designated changes of the split of one-way tours and round-trips over time, assumptions for the resulting distances to be traveled need to be made, too. Here, a main difference to detailed discrete event models as described in the following section can be seen, where the distance travelled is a result of integrated planning processes.

Alglawe et al. (2019) present a general framework for the behavior of all quality cost factors within the supply chain (SC). The proposed cost of quality (COQ) model uses a SD approach. In addition, Villa et al. (2015) develop an SD model that already analyzes the decisions and interdependencies between customers, retailers, and suppliers from an economic research perspective. Thaller et al. (2017) present a specific application of SD in urban logistics operations. Furthermore, La Torre et al. (2018) show a SD model that examined customer behavior from a last-mile perspective.

5.3.2 Discrete Event Simulation Models

Discrete event simulation can be considered the tool of choice if a multitude of detailed rules and algorithms have to be respected (see Sect. 1.2.2). The major challenges in the retail applications are twofold: Retail system simulation requires (i) a very detailed model of the material flow control and (ii) mechanisms to measure energy data and then to assign them to the emission drivers in order to identify improvement potentials (Rabe and Goldsman 2019).

With respect to the control strategies, retail distribution systems are characterized by several levels of sophisticated control systems, e.g., for inventory management, replenishment, warehouse management, tour planning, and many others, which are essential for the behavior of the system. Therefore, simulation models have to mimic such control systems on a rather high level of detail, much more than it can be considered usual in many other logistics tasks. In the specific case, further features can be found like mixed pallets, multi-drop delivery (requiring suitable tour planning algorithms), real-world order data handling, or specific cost calculation mechanisms such as freight cost matrices (Rabe et al. 2015).

Measuring energy data in a global view in retail can typically be conducted monitoring the major drivers, which are transport means and warehouses. Energy consumption of transport means is quite well researched (see also Chap. 4) and not discussed here. For warehouses, the major energy consumption raises from temperature regulation, which can be heating in winter and—much more relevant—cooling and refrigerating. Here, the total energy can normally be calculated rather easily, as it hardly depends on the logistics behavior. Therefore, the major challenge is often not the determination of the global energy consumption, but the assignment to the cost drivers (see Sect. 5.3.3).

Transportation processes Transport planning, i.e., solving a capacitated vehicle routing problem (VRP), needs to be an integral part of respective simulation models for retail distribution networks. Here, numerous restrictions are to be taken into account:

- Modeling fleets of trucks (of different types, including sleeping cabins, as these can have a significant impact on possible tour lengths)
- Truck capacity: Weight, volume, computation of detailed storage plans
- Maximum driving time, pauses, multi-day-trip restrictions
- Time windows for the delivery at stores or customers (e.g., inner-city restrictions),
- Truck type restrictions (e.g., no trailer allowed in certain urban areas, which makes it necessary to model tours that might include decoupling of a trailer ("outside of town") and picking it up again after serving a customer or store

In this context, the model validation is often problematic, since the number of tours and the distance travelled must be consistent with reality in order to obtain valid results. In order to obtain a valid model, also modeling the road network including possible traffic jams with rules for late arrivals (re-planning) might be necessary. With respect to results considering the energy consumption, the maximum speed for road segments should also be part of the model, as the energy consumption highly depends on the speed of the vehicles.

Within the simulation, daily tours often need to be computed to obtain valid overall distances and road segments to be taken due to stochastic demands with varying customers to be visited. In this context, simulation tools should offer the possibility to define respective problem formulations for the underlying VRPs and different heuristic approaches for solving dynamically occurring problem instances within each simulation run. Here, a possible process extension might be the decoupling of trailers before entering the city. A further extension is the planning of pickup-and-delivery tours, where a truck not only visits customers (or stores), but also picks up goods from suppliers on the same tour. In this context, Fig. 5.2 shows four different process types. A rather simple approach is to plan only direct tours from the suppliers to a central warehouse, and tours visiting several customers or stores from there (a). Introducing regional warehouses might lead to further direct tours from a central warehouse to several regional warehouses (b). An extension might be to also plan tours for the replenishment of the regional warehouses (c) or to further include suppliers on such tours, leading to combined pickup-and-delivery tours (d). Other combinations or process variants are, of course, possible, especially considering reverse logistics. For some transport relations, also multi-modal systems might be relevant. All these variants can have a massive impact on the energy consumption of the system, which is often not easy to forecast (see also the application case in Sect. 5.4.3).

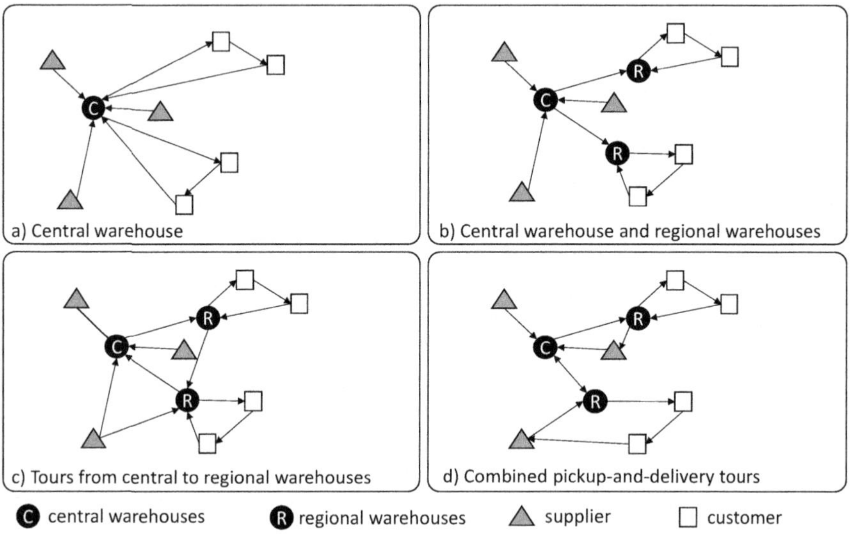

Fig. 5.2 Process types of retail supply

Data basis Modeling retail distribution systems is usually highly data intensive, because all relevant order and transport processes of the chosen excerpt of an overall retail network have to be modeled (Rabe et al. 2013; example given in Sect. 5.4.3):

- Energy consumption and CO_2 emission for different means of transport in case of an own fleet (total costs)
- Respective values given by the transport provider in relation to the distance travelled
- Order data (real world)
- Detailed road network, e.g., from *OpenStreetMap* or *GoogleMaps* with typical, daytime-dependent travel times between the respective locations, as the fuel consumption and energy consumption also depend on the speed and the overall travel time
- Modeling different temperature zones for storage and transport that affect the emission of CO_2

Therefore, retail simulation models require a very sophisticated data handling concept with clear mechanisms to achieve data quality and reliability of the results.

Simulation Tools These mechanisms have to be implemented in suitable tools. Several simulation tools may be used to model retail systems. Kuhl and Zhou (2009) present a concept for a simulation-based sustainability toolkit and a prototype being developed for modeling and simulating sustainability aspects of logistics and transportation systems using Arena (Rockwell Automation 2022). The main focus is the simulation modeling and statistics collection of performance indicators for energy consumption and emissions, as they relate to sustainability in addition to traditional productivity measures. In particular, the modeling and statistics collection involve determining the objects that consume energy and generate emissions and the events in the systems that cause the objects to start, stop, or change the consumption and emission rates. Another tool is SimChain (SimPlan AG 2022), which forms a shell around the commercial system Plant Simulation. SimChain has the advantage that it natively offers very sophisticated models of control systems on different levels, like inventory management, replenishment and supply decisions, or tour planning, including the handling of mixed pallets and many other detailed features that are characteristic for retail distribution. Therefore, the modeler can focus on the really specific challenges of the current study case and just select among the available control models. Within the European project E-SAVE, this tool has been enriched with a number of mechanisms to determine energy consumption and to assign this to consumption drivers.

5.3.3 *Performance Indicators*

There are two challenges for the definition of key performance indicators (KPIs) in retail system simulation (Rabe and Goldsman 2019). The first is to measure the global consumption. This can be conducted quite straight-forward, if the company runs their own fleet of trucks and their own warehouses. There are reasonable estimates for the energy consumption of different means of transport that can be used to realistically estimate the consumption of a specific system, if the actually travelled routes from the simulation results are known. For warehouses, heating and cooling efforts can often just be handled independently of the logistics and are, thus, easy to measure. A more difficult situation arises, if 3rd party logistics (3PL) enterprises are contracted, because their data are in most cases not accessible. In such cases, modeling the whole system is obviously not reasonable. Nevertheless, in order to optimize the specific system, transports provided by a third party cannot be just neglected. Therefore, estimations about their figures are required, e.g., the additional load that they can arrange for other clients or the tours that they can accept to avoid empty returns. A related problem arises if the routes of customers to shops or pickup locations have to be modeled (see Sect. 5.4.1).

The major issue, however, is the identification of the energy consumption drivers. For this purpose, the global consumption values need to be broken down and assigned to more detailed measures following specific distribution rules. Such rules can be quite subjective, but have a significant impact on the results.

Transport energy KPI For transport means, there are usual assumptions to estimate the fuel and, thus, energy consumption of a truck or another means of transport (see Sect. 5.4). It might be necessary to apply correction factors with respect to the logistics case. If the truck may take additional loads, e.g., in the case of a 3PL, a factor should be applied that reduces the energy consumption charged to the retail process. On the other hand, if there are potential empty returns of the trucks to the supplier, a corrective factor should be added that enlarges the energy consumption following the empty return drive. In reality, several further aspects will appear, which influence the specific energy consumption in transport processes. Modeling examples could consider traffic jams as a function of the week day and the time of the day, analyze the shape of the geographic environments (hills or plains), and not to forget the behavior of the driver, which has been proven to be a very significant factor. Also, the number of traffic lights on the route might be important, as it contributes to the acceleration activities, which cost extra energy. Demir et al. (2014) summarize these factors under the five categories vehicle, environment, traffic, driver, and operations. Frequently, the driver's behavior is neglected, because it is rather difficult to model. However, exactly this behavior is especially significant (Demir et al. 2014) and can overrule many of the other effects. Therefore, trying to calculate all the other categories can be quite questionable if the driver effects are uncertain. Another point are the structures of the road system, which could be expected to be different for specific types or topologies. Surprisingly, Rabe et al. (2020) have proven that the driving times and distances within different cities do hardly depend on the topology, but can be excellently estimated by a Euclidian distance with a correction factor of 1.234.

The issue comes up when these energy consumptions need to be assigned to specific goods or groups of goods, specific suppliers, specific retail customers, specific regions, or following any other criteria that might help the managers in their optimization tasks. Splitting up the energy consumption of a truck on its load is easy for homogeneous loads, but becomes a challenge if there are quite different goods in the transport. Typical assumptions use the ratio of weight or the ratio of volume, or even a complicated algorithmic combination of both. Just as a sample, if a truck already has a load of quite heavy goods, the forwarder might decide to pack additional light goods on the same truck. Actually, in this case the light goods will be transported without any relevant additional energy effort—but how can this realistically be represented in the evaluation? The authors have to admit that there are still a number of open questions, even in research.

Warehousing KPI The energy in warehouses is normally a complete overhead calculation. There are costs for heating and cooling, lighting, and local transport means. Often, the energy consumption of the local handling can be neglected against the massive costs of cooling. Analogous to transport, a corrective factor has to be applied if the warehouse is not exclusively used, where such common use is mostly quite well negotiated and, thus, the ratio of use can be well estimated.

Again, the issue comes when assigning energy to goods. The basic idea is to split the warehouse logistics in small intervals, which are defined by any goods entering or leaving the warehouse or the warehouse zone under consideration. For each of these little intervals, the inventory can be determined by simulation, and the energy consumption per time unit of the warehouse can be calculated and put into relation to the different types of goods. Again, there are quite subjective decisions. The ration could be calculated, e.g., in terms of weight, volume, pieces, pallets, and many other criteria, as well as their combinations. A quite detailed description of these approaches is given by Rabe and Goldsman (2019).

5.4 Applications

In this section, three examples for retail distribution systems are presented, each addressing a different aggregation level of modeling. The first application deals with a core question for a central retail process, the comparison of home deliveries and distributed locations for customers to pick up their orders with automated parcel lockers. Here, a rather high level modeling approach is chosen. The second application addresses issues of reverse logistics, i.e., the simulation of product returns, which have drawn quite some attention within retail processes lately. The third example presents a study for a European food enterprise analyzing different distribution structures with respect to the energy consumption. Further examples for applications in the field of retail distribution systems are presented also in Chap. 4, addressing last-mile parcel deliveries (Sect. 4.4.1), closely related to the first application in this section, and e-grocery deliveries in a Spanish city (Sect. 4.4.2).

5.4.1 Fuel Consumption for Delivery to Parcel Lockers Versus Home Deliveries

E-commerce opens up a new distribution channel for manufacturers, turning them into retailers. Due to more fast and hassle-free delivery, the number of parcels being shipped has increased rapidly, but the revenue per single delivery has declined while the number of delivery locations continues to grow. In a highly competitive sector, this puts additional pressure on limited margins. The recent pandemic crisis has acted as a catalyst, dramatically increasing the speed of change. Along with the COVID-19 pandemic, online shopping has increased significantly in many categories. These habits appear to be continuing, as consumers express their plans to continue shopping online after the COVID-19 crisis. The categories where expected growth in online shoppers exceeds 35% include essentials such as over-the-counter medicine, groceries, household goods and personal care products. In other categories, such as skin care and makeup, clothing, and jewelry, customer growth is expected to exceed 15% (IPC 2020). Customer behavior drives much of the last-mile costs, such as missed deliveries and returns. The last-mile problem encompasses one of the most costly and environmentally damaging segments of the retail supply chain, where companies deliver goods to end customers (Brown and Guiffrida 2014). The recent trend toward green supply chains and social and environmental responsibility has led to many new green initiatives. One business strategy of retailers that is becoming increasingly popular is to offer deliveries via an Automated Parcel Locker (APL) as an alternative to home delivery. The authors consider the use of APLs such as 'pack-stations' or locker boxes as one of the most promising initiatives to improve urban logistics activities (Boudoin et al. 2013). APLs have electronic locks with variable opening codes and can be used by different consumers whenever it is convenient for them. The APL combines multiple lockers located in homes, workplaces, train

(a) (b)

Fig. 5.3 Illustration of current APLs by **a** DHL (Beemelmanns 2016) and **b** Amazon locker (Post&Parcel 2015)

stations, etc. The costs of delivery through APLs are lower than home delivery and the risk of missed delivery is avoided. Some studies confirm that online shoppers will use APLs more frequently in the future (Moroz and Polkowski 2016).

APLs are found all over the world. Boudoin et al. (2013) and Zurel et al. (2018) have provided a general overview of the different experiences. For example, German 'packstations' have been in operation since 2001, with Deutsche Post DHL Group starting this business 20 years ago. The company currently operates more than 9,300 APLs and plans to increase this number to 12,000 by 2023 (DHL 2022; Last Mile Prophets 2022). The roll-out of French APLs began in 2014, and by the end of 2015, 200 APLs were operating in high-traffic areas of the five largest French cities (Paris and the Paris region, Lyon, Marseille, and Bordeaux). Figure 5.3 shows examples of APLs currently operated by DHL in Germany (a) and Amazon in France (b).

Verlinde et al. (2018) note that APLs have several advantages over home delivery: less traffic in downtown areas, no double-parking in front of customers' homes, fewer failed home deliveries, time savings, fewer kilometers, fewer stops, out-of-hours deliveries, and lower costs for e-retailers and delivery companies. Environmental benefits include reduced pollutant emissions and noise through the potential reduction of delivery vehicles in the city. Social benefits are expected in the form of improved quality of life. E-customers are free to choose the delivery time (24/7 availability) and the most convenient APL location to pick up or ship their parcels. In this example, a comprehensive comparison of the fuel consumption generated by traditional home delivery compared to delivery via APLs is given.

The use case in this section takes the city of Dortmund, which is located in the state of North Rhine-Westphalia, Germany, as a study case. With about 600,000 inhabitants, it is the seventh largest city in Germany. The city is divided into 62 districts. Figure 5.4 shows the map of the city of Dortmund and the distances between the distribution center and demand points (districts).

For determining the total *demand per district*, the data of the number of parcels for the city of Dortmund following Rabe et al. have been used (2021). In their paper, the authors determined the 36-month performance of parameters such as APL users

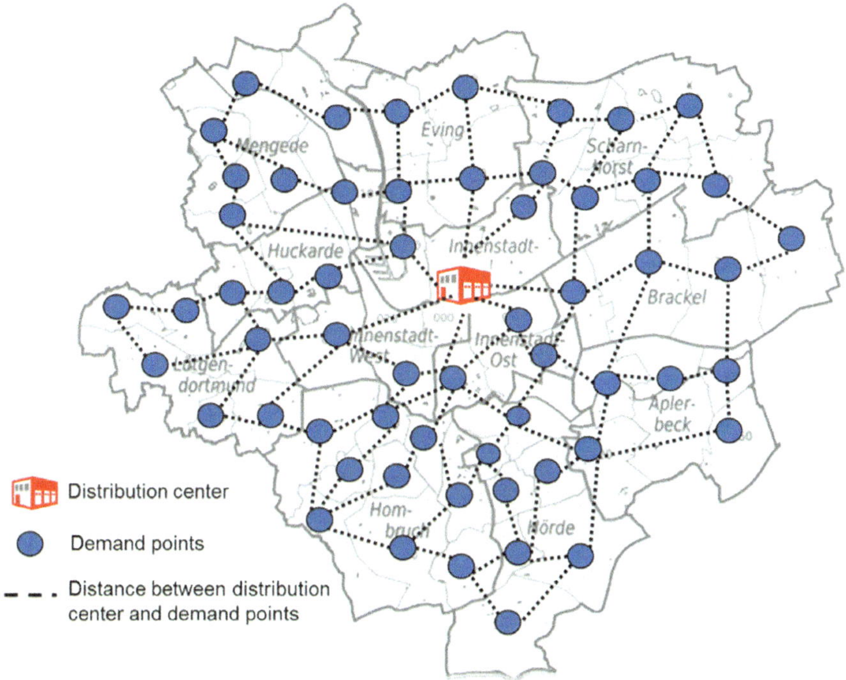

Fig. 5.4 Illustration of the districts in the city of Dortmund and the distances between distribution center and demand points (districts)

and number of deliveries using a SD model. In the same way, three scenarios S1, S2, and S3 were considered for 60%, 70%, and 80% of e-shopper rate, with an actual current e-shopper rate around 75% (Müller&Müller Consulting 2021). The experiments are based on Dortmund population data from December 2021 (IT.NRW 2021). The results for the number of parcels (units) for S1 increase from 314,000 in period 1 to 475,000 in period 36, for S2 from 366,000 to 550,000, and for S3 from 420,000 to 630,000, each period representing a month. Figure 5.5 illustrates the comparison of the scenarios for the number of parcels.

The *total distance for the delivery of parcels* is made up of two components: One is the distance from the distribution center to the customer, known as the "line-haul". This part of the route is traditionally calculated by solving the Capacitated Vehicle Routing Problem (CVRP). The other part is related to the distance between customers, which is relevant for the home delivery scenario. In order to estimate the distances over the entire district, Daganzo (1984) proposes the following intuitive formula for calculating the length of the line haul when the distribution center is located outside the customer's area:

$$d_{lh} = \frac{2rn}{Q}$$

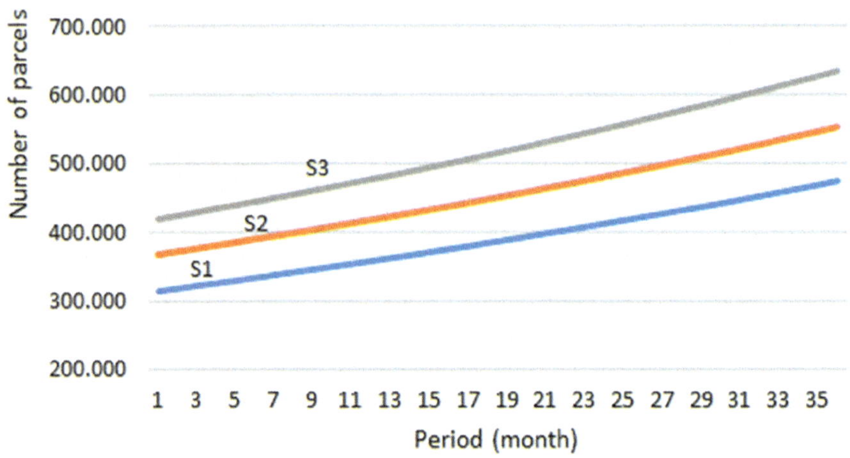

Fig. 5.5 Number of parcels for scenarios S1, S2, and S3 (Rabe et al. 2021)

with r being the distance between the distribution center and the area, n being the number of parcels to be served and Q being the capacity of each delivery van.

For the second component, approximate values for computing the distance between customers can be found in Beardwood et al. (1959). The authors show that the distance to travel between a set of n points in area A converges to $k\sqrt{nA}$, where A is the area containing the customers expressed in square kilometers. The constant term was estimated at $k = 0.765$, assuming compact and convex shapes for the areas where the tour is circumscribed (Stein 1978; Figliozzi 2009; Cárdenas et al. 2017).

The vehicle kilometers travelled (VKT) were calculated for all three scenarios. For the calculation of r, the distance between the distribution center and the districts has been estimated using a web mapping service. For the calculation of n, the data from Rabe et al. (2021) have been used again. The main road network of a district could be modeled for solving instances of the CVRP again to obtain more realistic results or validate such assumptions (see Sect. 5.3.2).

The van capacity Q was determined as 250 parcels per trip. The monthly demand is distributed equally between 24 days, such that the number of tours per district is the same for each day of the month. For each tour, the number of parcels is computed next, and finally the length of each tour is approximated following the approach described above. Public information was used to determine the area A of each district.

For the case of APLs, it is assumed that at least one APL operates with a capacity of 250 parcels per day for each district, i.e., one trip is required for every 250 parcels. In the case that more than one APL is located within a district, the VKT to each APL is approximated as twice the distance from the DC to the center of the respective district.

In the case of home deliveries after 36 months, the results for the VKT show a wide range from about 75,000 km in S1 to almost 130,000 km in S3. The VKT results

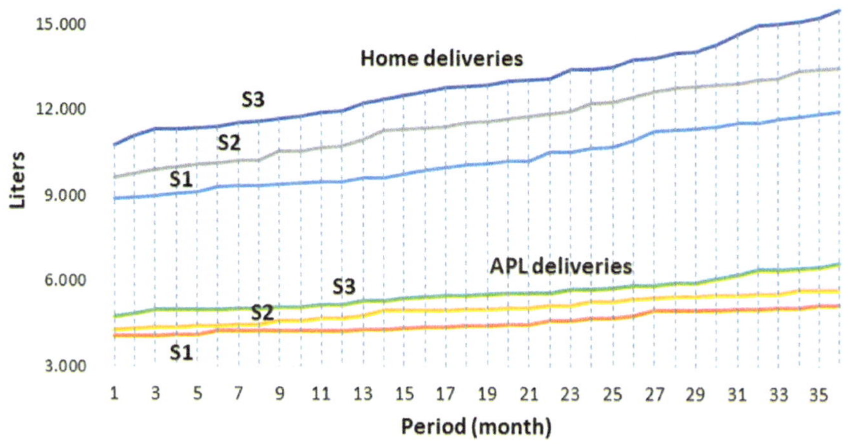

Fig. 5.6 Comparison of fuel consumption of home deliveries and deliveries using APLs for scenarios S1, S2, and S3

in the same period in case of APL usage range from 34,500 km in S1 to 56,000 km in S3.

Based on previous studies, the calculation of total *fuel consumption* is a crucial part of the approach reported here (Liu and Helfand 2009; Demir et al. 2011; Aksoy et al. 2014). To obtain realistic results, the fuel consumption is calculated considering the technical data of one of the most popular commercial vehicles for last-mile transport, the Mercedes-Benz Sprinter Cargo van, which uses the same engine as the passenger version. According to the technical data, the fuel consumption is 11.9 l/100 km in city traffic (cf. DAT 2020 for more information on the official fuel consumption of the Sprinter Cargo van). Figure 5.6 shows the fuel consumption results in the three scenarios for home deliveries and also for the use of APLs.

The fuel consumption for the case of home deliveries after 36 months is showing a wide range from about 9,000 L in S1 to more than 15,000 L in S3. The fuel consumption results for the same period when APLs are used ranges from about 4,000 L in S1 to more than 6,600 L in S3.

From the results it can be concluded that fuel consumption is directly correlated with vehicle type and distance travelled. The fuel consumption was calculated considering the technical specifications of the vehicle and the VKT from the distribution center to the delivery of the parcels to the districts. The results presented in this section show, on the one hand, that the use of APLs reduces VKT and, thus, fuel consumption by 56% compared to home delivery.

However, it should be considered that the retail systems are not modeled completely in the second case, and further assumptions need to be made on how customers travel to the APLs. Three different customer travel types are differentiated:

- C1: Customers could use their own car for a single tour to the APL and back home (worst case)

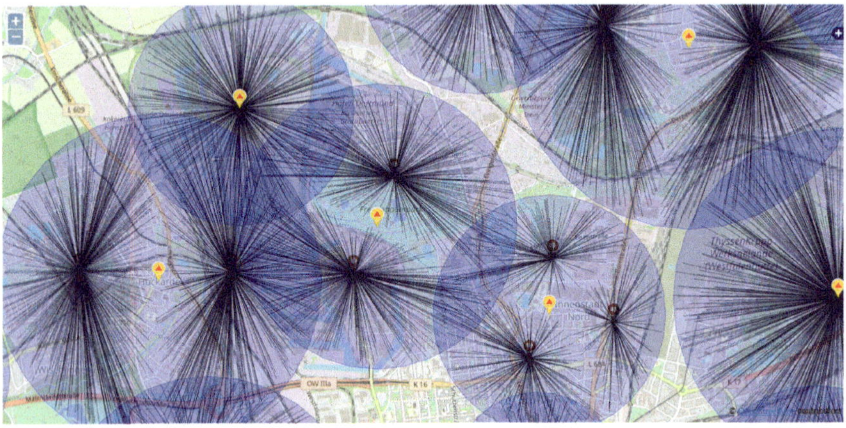

Fig. 5.7 Randomly generated customer locations and assignments to APLs within each district (OpenStreetMap)

- C2: Customers could make a stop at the APL with their car or using public transport on a different round trip, adding only a small degree of extra fuel consumption for picking up the parcel at the APL
- C3: Customers close to an APL might walk to the APL or take a bike, causing no further fuel consumption (best case)

For a rough estimation of the VKT of customers to the APLs, a simple model has been set up, where a fixed number of customer locations (based on the area and the population of the district) was randomly generated. The APLs' geographical locations have been defined in a way that the sum of the distances to the customer locations becomes minimal (assigning each customer to one of the APLs). An exempt of the map for scenario S3 (with a maximum of three APLs per district) is given in Fig. 5.7.

For each parcel, a customer is randomly selected and then the customer travel type is determined. For this purpose, a simple table for parameterization is used, which allows for giving the respective ratios for C1, C2, and C3 depending on the distance to the APL. In case the distance is less than 300 m, the assumption is that 100% of the customers walk or take a bike to the APL. If it is less than 1.5 km, the ratio is set to 50%, for longer distances to only 10%. In case a car is taken, the assumption is that 50% of the pickups are conducted on a direct trip to the APL, i.e., the VKT are twice the distance from the selected customer location to the APL; otherwise, 30% of the distance from the customer location to the APL are added to the customer tour as extra km. The fuel consumption for customer vehicles is selected as 5.6 l/100 km in city traffic on average.

For all three scenarios, the VKT of the customers and for serving the APLs outweigh the VKT for home deliveries by far, such that also the total fuel consumption is lower for home deliveries. Figure 5.8 represents these results for scenario S3.

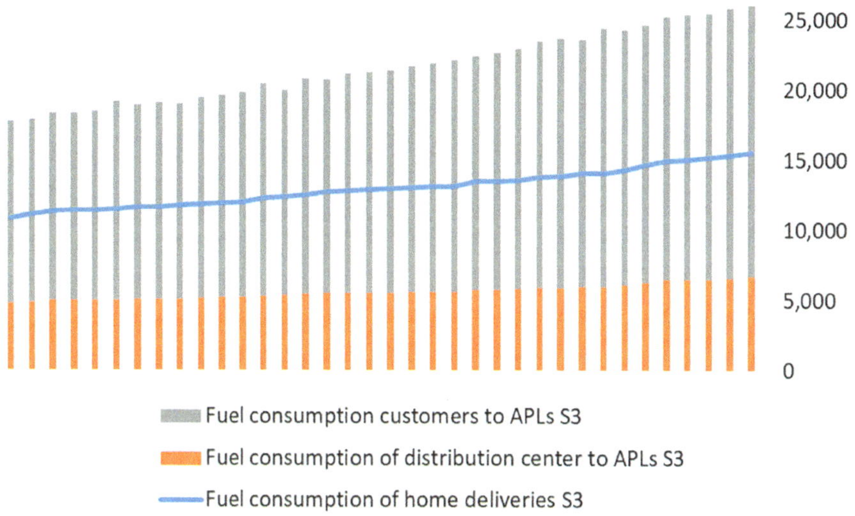

Fig. 5.8 Comparison of fuel consumption for home deliveries and APLs including customer pickups of parcels for scenario S3

Of course, APLs will become more attractive with an increasing number of APLs per district, such that customers will mainly walk or take a bike to the respective APL due to the closer distances.

5.4.2 *Reverse Logistics: Simulation of Product Returns*

Retail returns are a significant component of the costs of doing business. According to National Retail Federation and Appriss Retail (2022), in the U.S. retail returns accounted for $ 761 billion in lost sales in 2021. This is a return rate of 16.6% of total sales. In addition to the direct costs of lost sales, companies incur the costs of processing the returned items which include time, energy, labor, and other resources associated with reconditioning and remanufacturing, repackaging, recycling, or disposing of the item depending on the returned condition.

In addition to direct returns to retail stores or ecommerce retailers, companies are moving toward taking responsibility for the products that they produce (in some cases voluntary and in other cases compulsory) at the end of their useful life (as determined by the consumer) (Sherratt 2013). Industries such as the automotive parts industry have been doing this for decades. Others such as electronics and cell phone industries have more recently ramped up their efforts in reclaiming used products. Regardless of whether the product is a "new" product return or an end-of-useful-life return, systems are required to collect, disposition, and process the items.

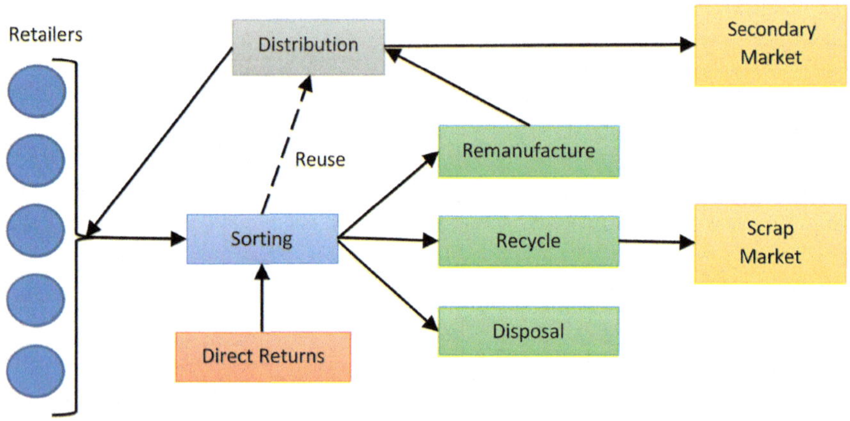

Fig. 5.9 Retail reverse logistics network flow

To address the issue of returns, Yanikara and Kuhl (2015) present a simulation framework for comparing alternative reverse logistic network configurations based on performance measures including productivity, energy usage, and environmental sustainability metrics.

The functionality of a typical reverse logistics network for retail products is shown in Fig. 5.9. Products returned or collected at the retailers are shipped directly to a sorting facility. A sorting process dispositions the products into four basic categories of reuse, remanufacture, recycle, or dispose. New (unused) products in their original packaging may be directly reused and distributed back to the retailers to be resold. Products that can be reconditioned, repaired, repackaged, etc. are sent to remanufacturing and sent for distribution either back to the retailers or to a secondary market. Products that are beyond repair may be sent to recycling or disposal. Reclaimed material from the recycled products can be sold in the scrap market.

Although the reverse logistics process seems straight forward, there are many decisions that need to be made to design a sustainable system. One of the challenges of reverse logistics system design is the random nature of the process including return volumes, product types, product condition, and processing time variability, among others. As a result, simulation is an excellent tool for analyzing reverse logistics systems in a way that can respect the uncertainties, capture the interdependencies, and measure the trade-offs among system alternatives.

One application of simulation to a reverse logistics network is determining the appropriate configuration of collection, sorting, and reprocessing. Figure 5.10 illustrates a group of twelve retailers that collect returns of multiple products. The challenge is to design an efficient network to process the returns. In Fig. 5.10, there are two potential system configurations. In Fig. 5.10a, products collected at the retailers are sent to a centralized sorting center with a nearby processing (remanufacturing and recycling) center. In Fig. 5.10b, the sorting task is distributed between two sorting

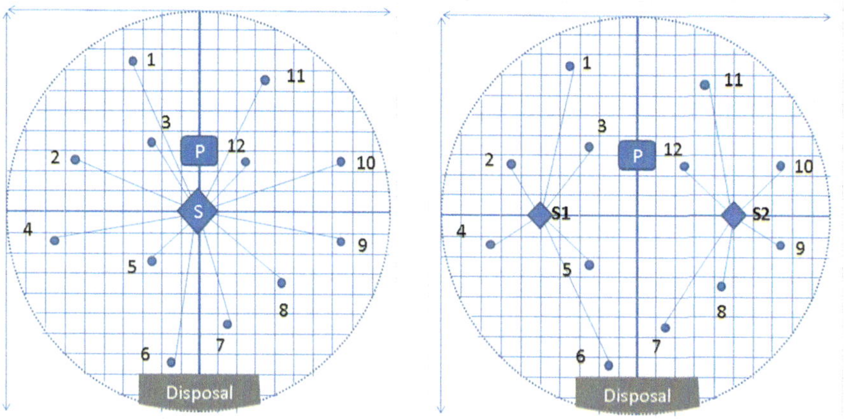

(a) (b)

Fig. 5.10 Two potential reverse logistics system configurations

centers and one centralized processing center. These are just two of the many possible system configurations that could be considered.

Using simulation, the dynamic behavior of alternative systems can be analyzed based on a set of performance metrics related to productivity, energy consumption, value of resale products and recovered material, and emissions. Productivity measures include sorting and processing costs, inventory costs, etc. Energy consumption includes the electricity and fuel for processing and transportation. The monetary value of the remanufactured products and material recovered through recycling can also be collected. Finally, the emissions generated from processing and transportation such as CO_2, CH_4, and N_2O emissions can be collected from the simulation. System alternatives can be compared by evaluating the trade-offs among the various performance measures. Since the various performance measures are not in the same measurement units, another option is to use a weighted factor comparison to generate a score on which the system variants could compared.

5.4.3 Distribution Structures in the Food Sector

The third application presented here is a study at a major European food enterprise, which operates European-wide with different product categories and production sites in several countries. The study has been focused on two types of goods, namely noodles and related sauces ("pasta") and crispbread, and furthermore on a single market (Germany) and one large supermarket chain as the customer. The supermarket chain supplies the retail shops—which run under several different brand names—from regional distribution centers. Consequently, the customers (or sinks) of the distribution from the food company's point of view are these distribution centers.

In the as-is situation, both distribution systems have been independent. Bread was send with trucks from the WASA factory in Celle (northern Germany) to the regional distribution centers. Pasta was sent with a different fleet of trucks from the Barilla factory in the north of Italy either to the supermarket's regional centers or to a consolidation center in the southern part of Germany, where it was put on stock and commissioned for the regional centers of the supermarket on short-term demand. The suspicion has been that this strategy causes unnecessary waste of resources by running partially empty trucks, correlated with additional emissions. Thus, the question was: Would it be economically and environmentally reasonable to consolidate all goods (pasta and bread) in one consolidation center in Germany, and then try to operate full-load trucks from there to the regional centers? The saving by achieving better truck loading factors was obvious, but the consolidation centers would have either raised the need to transfer all bread products there, or else to first convey all pasta products to the north of Germany, even if half of it would later be ordered from the south. Therefore, the question required a careful analysis, which has been conducted in the framework of a European-supported project (Rabe et al. 2015).

Based on the real data from the IT systems of the manufacturer from one complete year, all products have been filtered that belong to the addressed product groups and are purchased by the selected supermarket chain. The network included 586 locations. Within the one-year time period under study, the WASA network had 8,006 orders and Barilla received 13,218 orders. A number of different scenarios was defined, with a consolidation center in the south or the north, or even both (see examples in Fig. 5.11).

The study was based on the supply chain simulation tool SimChain, which in this enhanced version has been developed during the conduction of the mentioned

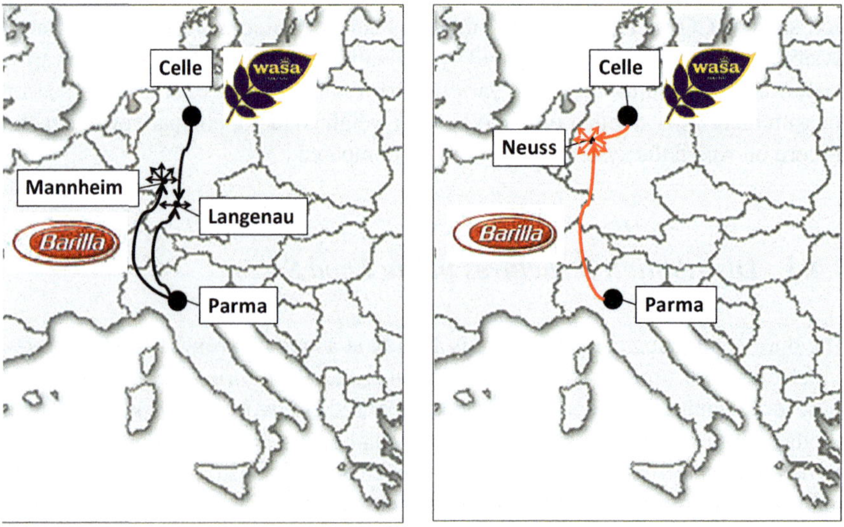

Fig. 5.11 Two examples for potential location of consolidation centres (Rabe et al. 2015)

European project and enriched with a suitable data model to calculate and assign energy consumption (Rabe et al. 2013). Unexpectedly, the results have shown that the as-is situation provides the best results in terms of all categories costs, emissions, and service levels. Actually, with the variants energy consumption would be about 9% higher than currently. The worst results occur when all goods pass through only one distribution node. Therefore, it can be concluded that the added transport of goods to the consolidation centers is overcompensating the benefits that are achieved by the higher truck loads. Details of the considered scenarios and more detailed results can be found at Rabe et al. (2015).

5.5 Conclusions and Outlook

In this chapter, an overview has been provided of the simulation of retail distribution systems along with three applications. Looking at the literature, the explicit consideration of energy consumption and CO_2 emissions has become of higher interest in the last years, with the idea of green supply chains and a stronger focus on reverse logistics appearing in more and more publications. A broad application of such models, intending to reduce emissions and energy consumption, is still on its way, especially when analyzing large retail systems. Here, system dynamics can be used for rather strategic decisions looking at markets and long-term developments, while discrete event simulation is an excellent tool to analyze underlying processes explicitly modeling transports, warehouses (with different cooling zones), and order policies.

Looking at future applications, simulation tools should be further enhanced by including energy-relevant properties of system elements, such as resources and buildings (with respect to heating, lighting, and cooling). For retail systems, a highly interesting and important field for future research is modeling the movements of customers to shops and pickup-locations, which are needed to compare the main key figures for energy consumption and emissions with systems containing only home deliveries.

References

Abbas KA, Bell MGH (1994) System dynamics applicability to transportation modeling. Transp Res Part A 28(5):373–400

Aksoy A, Küçükoğlu İ, Ene S, Öztürk N (2014) Integrated emission and fuel consumption calculation model for green supply chain management. Procedia Soc Behav Sci 109(2):1106–1109

Alglawe A, Schiffauerova A, Kuzgunkaya O (2019) Analysing the cost of quality within a supply chain using system dynamics approach. Total Qual Manag Bus Excell 30(15–16):1630–1653

Angerhofer BJ, Angelides MC (2000) System dynamics modelling in supply chain management: Research review. In: Joines JA, Barton RR, Kang K, Fishwick PA (eds) Proceedings of the 2000 Winter Simulation Cconference. IEEE, Piscataway, NJ, pp 342–351

Auf der Landwehr M, Trott M, von Viebahn C (2021) Environmental sustainability as food for thought! Simulation-based assessment of fulfillment strategies in the e-grocery sector. In: Kim S, Feng B, Smith K, Masoud S, Zheng Z, Szabo C, Loper M (eds) Proceedings of the 2021 Winter Simulation Conference. IEEE, Piscataway, NJ, pp 1–12

Beardwood J, Halton JH, Hammersley JM (1959) The shortest path through many points. Math Proc Cambridge Philos Soc 55(4):299–327

Beck N, Rygl D (2015) Categorization of multiple channel retailing in multi-, cross-, and omni-channel retailing for retailers and retailing. J Retail Consum Serv 27:170–178

Beemelmanns L (2016) DHL-Packstation an der Prämienstraße wird erweitert. https://www.aachener-nachrichten.de/lokales/stolberg/dhl-packstation-an-der-praemienstrasse-wird-erweitert_aid-25082967. Accessed 27 Mar 2022

Boudoin D, Morel C, Gardat M (2013) Supply chains and urban logistics platforms. In: Gonzalez-Feliu J, Semet F Routhier J (eds) EcoProduction. Sustainable urban logistics: concepts, methods and information systems. Springer, New York, NY, pp 1–20

Brown J, Guiffrida A (2014) Carbon emissions comparison of last mile delivery versus customer pickup. Int J Log Res Appl 17(6):503–521

Cárdenas I, Beckers J, Vanelslander T (2017) E-commerce last-mile in Belgium: developing an external cost delivery index. Res Transp Bus Manag 24:123–129

Daganzo CF (1984) The distance traveled to visit n points with a maximum of c stops per vehicle: an analytic model and an application. Transp Sci 18(4):331–350

Das D, Dutta P (2013) A system dynamics framework for integrated reverse supply chain with three way recovery and product exchange policy. Comput Ind Eng 66(4):720–733

DAT (2020) Leitfaden über den Kraftstoffverbrauch, die CO_2-Emissionen und den Stromverbrauch neuer Personenkraftwagen. https://www.dat.de/co2/. Accessed 25 Mar 2022

Demir E, Bektaş T, Laporte G (2011) A comparative analysis of several vehicle emission models for road freight transport. Transp Res Part D 16(5):347–357

Demir E, Bektaş T, Laporte G (2014) A review of recent research on green road freight transportation. Eur J Oper Res 237(3):775–793

DHL (2022) DHL packstation. https://www.dhl.de/en/privatkunden/pakete-empfangen/an-einem-abholort-empfangen/packstation.html. Accessed 31 May 2022

Duthoit A (2021) European food retailers: the bitter digital aftertaste of the COVID-19 legacy. Allianz Research. https://www.allianz-trade.com/en_global/news-insights/economic-insights/European-food-retailers-The-bitter-digital-aftertaste-of-the-Covid-19-legacy.html. Accessed 22 May 2022

Figliozzi MA (2009) Planning approximations to the average length of vehicle routing problems with varying customer demands and routing constraints. Transp Res Record: J Transp Res Board 2009(1):1–8

Georgiadis P, Vlachos D, Iakovou E (2005) A system dynamics modeling framework for the strategic supply chain management of food chains. J Food Eng 70(3):351–364

Gutenschwager K, Arnold P (2020) Simulation von Lieferantennetzwerken: Grundlagen und Anwendungen bei der ZF Friedrichshafen AG. In: Mayer G, Pöge C, Spieckermann S, Wenzel S (eds) Ablaufsimulation in der Automobilindustrie. Springer, Berlin, pp 247–259

Hübner A, Holzapfel A, Kuhn H (2016) Distribution systems in omni-channel retailing. Bus Res 9:255–296

IPC (2020) Covid-19 and the long-term changes to e-commerce. https://www.ipc.be/services/markets-and-regulations/e-commerce-market-insights/e-commerce-articles/2020-10-covid-and-ecommerce. Accessed 31 May 2022

IT.NRW (2021) Bevölkerung in Nordrhein-Westfalen. https://www.it.nrw/statistik/eckdaten/bevoelkerung-nach-gemeinden-93051. Accessed 31 May 2022

Kazancoglu Y, Ekinci E, Mangla SK, Sezer MD, Kayikci Y (2021) Performance evaluation of reverse logistics in food supply chains in a circular economy using system dynamics. Bus Strateg Environ 30(1):71–91

Koetsier J (2022) E-commerce jumped 55 % during covid to hit $1.7 trillion. Forbes, Mar 15. https://www.forbes.com/sites/johnkoetsier/2022/03/15/pandemic-digital-spend-17-trillion/. Accessed 25 May 2022

Kuhl ME, Zhou X (2009) Sustainability toolkit for simulation-based logistics decisions. In: Rossetti MD, Hill RR, Johansson B, Dunkin A, Ingalls RG (eds) Proceedings of the 2009 Winter Simulation Conference. IEEE, Piscataway, NJ, pp 1466–1473

Last Mile Prophets (2022) Deutsche Post DHL's Packstation network—is 12,000 enough? https://lastmileprophets.com/deutsche-post-dhl-packstation-parcel-locker/. Accessed 31 May 2022

La Torre G, Gruchmann T, Kamath V, Melkonyan A, Krumme K (2018) A system dynamics-based simulation model to analyze consumers' behavior based on participatory systems mapping—a "last mile" perspective. In: Melkonyan A, Krumme K (eds) Innovative logistics services and sustainable lifestyles. Springer Science and Business Media, New York, NY, pp 165–194

Liu Y, Helfand GE (2009) The alternative motor fuels act, alternative fuel vehicles, and greenhouse gas emissions. Transp Res Part A 43(8):755–764

Moroz M, Polkowski Z (2016) The last mile issue and urban logistics: choosing parcel machines in the context of the ecological attitudes of the Y generation consumers purchasing online. Transp Res Procedia 16:378–393

Müller & Müller Consulting (2021) The e-commerce market in Germany. https://www.muellersconsulting.com/post/statistics-e-commerce-market-germany. Accessed 1 Jun 2022

National Retail Federation and Appriss Retail (2022) Consumer returns in the retail industry 2021. https://apprissretail.com/resource/2021-consumer-returns-in-the-retail-industry/. Accessed 4 Dec 2022

Post & Parcel (2015) Amazon reportedly installs first lockers in France. https://postandparcel.info/69851/news/amazon-reportedly-installs-first-lockers-in-france/. Accessed 25 Mar 2022

Rabe M, Goldsman D (2019) Decision making using simulation methods in sustainable transportation. In: Faulin J, Grasman SE, Juan AA, Hirsch P (eds) Sustainable transportation and smart logistics: decision-making models and solutions. Elsevier, Amsterdam, pp 305–333

Rabe M, Gutenschwager K, Fechteler T, Sari MU (2013) A data model for carbon footprint simulation in consumer goods supply chains. In: Pasupathy R, Kim S-H, Tolk A, Hill RR, Kuhl ME (eds) Proceedings of the 2013 Winter Simulation Conference. IEEE, Piscataway, NJ, pp 2677–2688

Rabe M, Sari MU, Fechteler T, Ruini LF (2015) Discrete event simulation as a strategic decision instrument for a CO_2- and cost-efficient distribution chain applied in the FMCG sector. Int J Adv Logistics 4(1):47–53

Rabe M, Klüter A, Raps J (2020) Evaluating different distance metrics for calculating distances of last mile deliveries in urban areas for integration into supply chain simulation. J Simul 14(1):41–52

Rabe M, Gonzalez-Feliu J, Chicaiza-Vaca J, Tordecilla RD (2021) Simulation-optimization approach for multi-period facility location problems with forecasted and random demands in a last-mile logistics application. Algorithms 14(2):41. https://doi.org/10.3390/a14020041

Rockwell Automation (2022) ARENA simulation software. https://www.rockwellautomation.com/en-us/products/software/arena-simulation.html. Accessed 20 May 2022

Sarkis J, Zhu Q, Lai K (2011) An organizational theoretic review of green supply chain management literature. Int J Prod Econ 130(1):1–15

Shepherd SP (2014) A review of system dynamics models applied in transportation. Transportmetrica B: Transp Dyn 2(2):83–105

Sherratt A (2013) Cradle to cradle. In: Idowu SO, Capaldi N, Zu L, Gupta AD (eds) Encyclopedia of corporate social responsibility. Springer, Berlin Heidelberg, pp 630–638

SimPlan AG (2022) SimChain. https://www.simchain.net/. Accessed 20 May 2022

Srivastava SK (2007) Green supply-chain management: a state-of-the-art literature review. Int J Manag Rev 9(1):53–80

Stein DM (1978) An asymptotic, probabilistic analysis of a routing problem. Math Oper Res 3(2):89–101

Susanty A, Puspitasari D, Purwaningsih R, Winarti RY (2016) Impact of green supply chain management practices on enterprise performance in furniture industry: a review for conceptual model. Adv Sci Lett 22(5–6):1096–1104

Thaller C, Niemann F, Dahmen B, Clausen U, Leerkamp B (2017) Describing and explaining urban freight transport by system dynamics. Transp Res Procedia 25:1075–1094

Verlinde S, Rojas C, Buldeo Rai H, Kin B, Macharis C (2018) E-consumers and their perception of automated parcel stations. In: Taniguchi E, Thompson R (eds) City logistics 3: towards sustainable and liveable cities. John Wiley and Sons, Hoboken, NJ, pp 147–160

Von Viebahn C, Auf der Landwehr M, Trott M (2020) The future of grocery shopping? A taxonomy-based approach to classify e-grocery fulfillment concepts. https://doi.org/10.30844/wi_2020_j2-viebahn. Accessed 24 May 2022

Villa S, Gonçalves P, Arango S (2015) Exploring retailers' ordering decisions under delays. Syst Dyn Rev 31(1–2):1–27

Yanikara FS, Kuhl ME (2015) A simulation framework for the comparison of reverse logistic network configurations. In: Yilmaz L, Chan WKV, Moon I, Roeder TMK, Macal C, Rossetti MD (eds) Proceedings of the 2015 Winter Simulation Conference. IEEE, Piscataway, NJ, pp 979–990

Zurel Ö, van Hoyweghen L, Braes S, Seghers A (2018) Parcel lockers, an answer to the pressure on the last mile delivery? In: Parcu PL, Brennan TJ, Glass V (eds) New business and regulatory strategies in the postal sector. Springer International, Cham, pp 299–312

Chapter 6
Perishables

Christian Fikar, Björn Johansson, Karsten Beyer, and Marvin Auf der Landwehr

Abstract Perishable goods such as fruits and vegetables require timely and accurate handling routines to ensure a high degree of product quality across all stages of the supply chain. Consequently, they constitute a fundamental business factor for organizations that needs to be managed in a delicate and prudent fashion. The perishability of products characterizes a challenging environment that requires dynamic planning and evaluation approaches to avoid or countervail the negative energetic impacts of inefficient operations. By providing a sophisticated conceptualization of the given system and its dynamic evolution over time, computer simulation serves as viable tool for analyzing and optimizing energy-related aspects of production and logistics systems for perishable items. This chapter reviews the current state of research for simulating energy-related aspects of perishable products and highlights common energy performance indicators such as food waste, emissions, and temperature. To outline contextual interdependencies and provide practical insights into the use of simulation to assess energy aspects of perishables, three use cases are presented. These cases elaborate on the energetic implication of a juice production plant in Sweden, the estimation of food quality losses in regional strawberry supply chains in Austria, and the energy and media consumption of a beverage bottling plant in Germany.

C. Fikar (✉)
Faculty VII for Life Sciences: Food, Nutrition and Health, Chair of Food Supply Chain Management, University of Bayreuth, Fritz-Hornschuch-Strasse 13, 95326 Kulmbach, Germany
e-mail: christian.fikar@uni-bayreuth.de

B. Johansson
Department of Industrial and Materials Science, Chalmers University of Technology, Hörsalsvägen 7A, 412 96 Gothenburg, Sweden
e-mail: Bjorn.Johansson@chalmers.se

K. Beyer
SimPlan AG—Dresden Branch, Gostritzer Straße 63, 01217 Dresden, Germany
e-mail: Karsten.Beyer@simplan.de

M. Auf der Landwehr
Fakultät Wirtschaft und Informatik, Institute for Applied Data Science Hanover, Hochschule Hannover, Ricklinger Stadtweg 120, 30459 Hannover, Germany
e-mail: marvin.auf-der-landwehr@hs-hannover.de

S. Wenzel et al. (eds.), *Energy-Related Material Flow Simulation in Production and Logistics*, https://doi.org/10.1007/978-3-031-34218-9_6

6.1 Introduction

Efficient and effective management of perishables continues to be a major success factor for manufacturers, suppliers, and retailers. Perishables refer to items that are subject to decay, ruin, or deconstruction and are present in various industries, including food, newspapers, medications, blood products, and organs (Jbira et al. 2018). Typically, these kinds of products account for a large proportion of sales, while serving as means for differentiation and creation of competitive advantage. Moreover, perishables tend to feature higher gross margins, which come along with increased costs for specific requirements in terms of labor, transportation, and packaging (Zhang et al. 2021). Since their utility to customers is highly limited based on their restricted lifetimes, which, in turn, entails short replacement cycles, they constitute both a major cost factor for organizations and a noticeable opportunity to increase sales revenues (Jbira et al. 2018).

To ensure high degrees of profitability, the production, storage, and transportation of perishable items needs to be supported by sophisticated machinery and infrastructure that meet energy-intensive requirements such as cooling or air humidity (Gharehyakheh et al. 2020). Storing and handling routines of perishable products have the dual goal of preserving the quality and safety of items, which is affected by the microbiological, physiological, biochemical, and physical activities occurring throughout the lifecycle, as well as fostering fast and reliable fulfillment processes. For the former objective, energy is used extensively to guarantee continuous refrigeration during production, storage and transportation and, consequently, slowing down the quality decay speed, contributing a major share of operational costs and significant emissions (Fan et al. 2021). For the latter goal, comprehensive supply and distribution networks as well as shipping capabilities are required, implicating considerable amounts of transportation-related emissions (De Keizer et al. 2015).

As synopsized in Fig. 6.1, the perishability of products and the uncertainty in environmental conditions experienced throughout the supply chain characterize a challenging environment that requires dynamic assessment and management approaches to avoid or countervail the negative effects of inefficient operations concerned with these kinds of products. Due to the fact that perishable products feature unique temporal (e.g., lifetime), operational (e.g., cooling) and spatial (e.g., storage) constraints as well as numerous decision variables such as inventories and lead times, their underlying production systems and supply chains are more complex compared to other industries (Van der Vorst et al. 2009). This complexity is not problematic per se, however, it makes it difficult to define exact analytical models and, therefore, requires a methodological approach that is capable of capturing uncertainties and stochasticity in multi-actor settings, while at the same time enabling to investigate multiple hypothetical configurations (Pirard et al. 2011). In this context, computer simulation has proven to be an effective tool for analyzing and optimizing energy-related aspects of production and logistics systems that are concerned with perishable products. The investigation of a system by means of simulation models enables decision makers to develop a sophisticated conceptualization and understanding of the

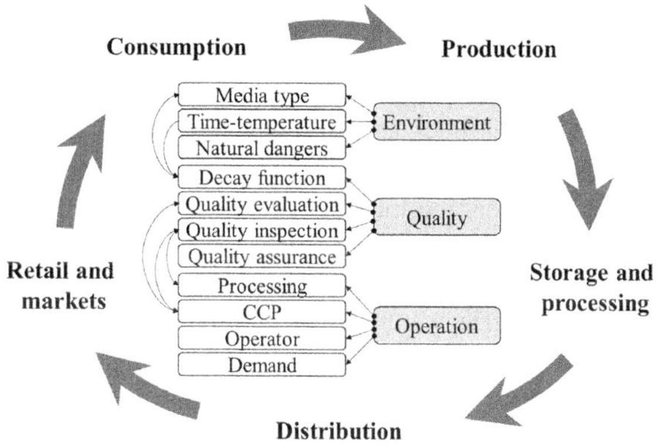

Fig. 6.1 Properties and challenges of supply chains dealing with perishables (CCP: Central Counterparty Clearing)

system and further allows for evaluating the dynamic nature of a phenomenon's evolution over time (McHaney 1991).

6.2 Simulation Background

Historically, simulation models have been employed in manifold contexts that involve perishable products. On the one hand, they can serve as tools for studying the behavior of different systems, including supply chains (e.g. Fan et al. 2021, Ketzenberg et al. 2015, Leithner and Fikar 2019), production plants (e.g., Akkerman et al. 2007; Ivanov and Rozhkov 2020; Kouki et al. 2013; Polotski et al. 2021), inventory management routines (e.g., Bottani et al. 2014; Zhang et al. 2021), and logistics networks (e.g. Haass et al. 2015, Lin et al. 2017, Lütjen et al. 2013). On the other hand, simulation technology is a viable instrument to identify suitable avenues for alternative process flows or system designs and optimize operational standards (e.g., Czerniak et al. 2021; Haijema et al. 2009; Noordhoek et al. 2018). In the context of food, scholars commonly model food quality changes and time temperature indicators to monitor the temperature conditions of food items throughout distribution on an individual basis (e.g., Psomas et al. 2011; Scott and Heldman 1990; Van der Vorst et al. 2009).

Simulation-based research on energy indicators generally involves a high amount of probabilistic system parameters (e.g., consumer demand or spoilage duration). To analyze or optimize such stochastic systems, simulation models capable of accommodating a multitude of stochastic and deterministic factors need to be formulated. According to Kelton and Sadowski (2009), a discrete event simulation (DES) is particularly useful to capture the inherent dynamics of probabilistic systems and optimize process-related workflows. The DES methodology requires the modeler to specify event triggers that determine discrete state changes during the execution of the simulation model, which is also referred to as time-advancing mechanism (Lin et al. 1996). Since it enables to track specific items and individual entities, this type of simulation is more appropriate than continuous simulation approaches (e.g., System

Dynamics) for analyzing energy-related performance indicators of production and logistics systems involving perishables. Moreover, Monte Carlo simulation is widely used by scientists and engineers to investigate decay effects of perishables during production, storage, and transportation phases (e.g., La Scalia et al. 2019).

Nevertheless, depending on a study's purpose, context of investigation, and problem space, the individual applicability of simulation methodologies can vary. While DES tools tend to stress logistics analysis and supply chain productivity, allowing for a substantial assessment level concerning systemic energy indicators such as electricity consumption, their focus on product quality and sustainability optimization is rather limited (Van der Vorst 2009). In turn agent-based modeling (ABM) and related simulation methods offer a natural way to model individual entities in multi-actor settings and implement multi-tier architectures (Behdani et al. 2012). Such a multi-tier architecture could consist of a social layer and a physical layer, whereby the social layer models each actor as an autonomous agent that makes decisions and interacts with other agents, and the physical layer incorporates objects that perform logistic activities, emit pollutants, and consume energy (Holmgren et al. 2012). Other simulation techniques include numerical simulation (e.g., Chatzidakis et al. 2004; Haass et al. 2015) and tool-specific building energy simulation (e.g., Burek and Nutter 2020).

Finally, scholars occasionally employ hybrid approaches that combine mathematical optimization and simulation into a single solution procedure to ensure a more holistic level of assessment (Hatami-Marbini et al. 2020). In a nutshell, hybrid methods allow for integrating both techniques in different forms and directions. For instance, local optimizations can be used to set parameters of a simulation model (e.g., Pirard et al. 2011), or a simulation model can be employed as means of solution evaluator in a search procedure (e.g., Ding et al. 2009; Melouk et al. 2013). Correspondingly, simulation and mathematical optimization can also be utilized in a sequential or iterative procedure, such as given in cases where simulation is used to provide feedback for re-optimization (e.g., Safaei et al. 2010; Sel and Bilgen 2014). Apart from coupling mathematical optimization and simulation, hybrid simulation can also refer to the combination of different modeling approaches or the interplay of simulation and machine learning (Mustafee et al. 2015). An overview on different categorizations of hybrid simulation along different dimensions can be found in Figueira and Almada-Lobo (2014) as well as Brailsford et al. (2019).

6.3 Energy Performance Indicators

Energy performance indicators can be assessed in a multiplicity of facets. Thereby, energy-related parameters in extant simulation-based research on perishable items can be clustered alongside three research streams, which correlate with the system under investigation, namely (1) supply chain management, (2) distribution, and (3) production. Table 6.1 provides an overview of the investigated energy-related performance indicators of selected studies for perishables in order to synopsize relevant indicators in scientific research.

In terms of the problem domain, supply chain management describes "the integrated planning, coordination, and control of all logistic business processes and activities in the supply chain to deliver superior customer value at less costs to the supply chain as a whole, while satisfying requirements of other stakeholders (e.g., the government or non-governmental organizations) in the wider context of the total supply chain network" (Van der Vorst et al. 2009, p. 6611). Energy performance indicators of supply chains related to perishable products include energy consumption, pollutant emissions (e.g., CO_2), as well as food loss and waste. While the former indicators constitute direct energy parameters, food waste features an indirect energy relation, since it can be synthesized with energy consumption rates for different categories of perishable items as well as the related production steps to calculate energy equivalents and losses (Cuéllar and Webber 2010). Similarly, greenhouse gas emissions can be associated with food losses and waste (Dong et al. 2021). In contrast to food losses, which refer to the quantity of edible food products that is lost across the supply chain, food waste describes the number of discarded products at the end of the supply chain and is, thus, only applicable to retailers and consumers (Hu et al. 2019). In summary, it can be said that food losses and waste entail a certain amount of energy and carbon that was wasted during agriculture, transport and production activities.

Research streams focusing on distribution predominantly examine constructs concerned with storage, handling, and transportation of goods to the customer, including vehicle routing, network performance, and vehicle design. Energy-related aspects in simulation-based research typically relate to greenhouse gas emissions and mileages, which accrue due to the required transportation activities. Moreover, energy consumption of refrigerants and cooling infrastructure are commonly researched, since distribution activities take a major share of the total energy consumption of refrigeration in the food industry (Wu et al. 2019).

Finally, production-related studies leverage simulation methods to investigate problems in the fields of production planning (cf. Hatami-Marbini et al. 2020), material flow management (Forster 2013), and building energy consumption (Santos et al. 2013). Prominent energy performance indicators are energy consumption in terms of electricity, water, steam, acid and gasses, as well as electrical power. Moreover, concerning indirect factors, machine utilization and food waste are commonly researched and can be used to draw conclusions about energy consumption or energy requirements (Hatami-Marbini et al. 2020).

Table 6.1 Energy-related performance indicators in simulation-based perishables research

Reference	Stream	Perishable items	Approach	Energy-related indicators	Other indicators
Aiello et al. (2012)	SCM	Food	MCS	Food waste[a]	Product lifetime, Product quality
Auf der Landwehr et al. (2021)	DIST	Food	ABM	CO_2 emissions, mileages[a]	None
Burek and Nutter (2020)	SCM	Food	Others	CO_2 emissions, fossil energy consumption, food waste[a]	Water scarcity
Chatzidakis et al. (2004)	DIST	Food	Others	Temperature, energy consumption	None
de Keizer et al. (2015)	DIST	Flowers	HS	Mileages[a]	Product quality, system costs
Dong and Miller (2021)	SCM	Food	MCS	CO_2 emissions, energy consumption	None
Fan et al. (2021)	SCM	Food	ABM	CO_2 emissions, energy consumption, food waste[a]	Product quality, system costs
Fikar (2018)	DIST	Food	ABM	Mileages[a], food waste[a]	Product quality
Forster (2013)	PROD	Beverages	DES	Energy consumption, water, steam, acid and gas consumption	Air pressure
Gharehyakheh et al. (2020)	DIST	Food	HS	CO_2 emissions	System costs, product lifetime
Gruler et al. (2017)	DIST	Garbage	HS	Mileages[a]	System costs
Haass et al. (2015)	DIST	Food	Others	CO_2 emissions, food waste[a], temperature,	Product quality
Hatami-Marbini et al. (2020)	PROD	Not specified	HS	Machine utilization[a], food waste[a]	System costs
Ketzenberg et al. (2018)	SCM	Food	Others	Temperature	System costs, product lifetime
La Scalia et al. (2019)	SCM	Food	MCS	CO_2 emissions, food waste[a]	Revenue

(continued)

Table 6.1 (continued)

Reference	Stream	Perishable items	Approach	Energy-related indicators	Other indicators
Leithner and Fikar (2019)	SCM	Food	DES	Food waste[a]	Product quality, service levels
Rijpkema et al. (2014)	SCM	Food	HS	Food waste[a]	System costs, product quality
Santos et al. (2013)	PROD	Food	Others	Energy consumption, electrical power	Product quantity
Shamsi et al. (2014)	DIST	Not specified	Others	CO_2 emissions, mileages[a]	System costs, empty trips
van der Vorst et al. (2009)	SCM	Food	DES	Energy consumption, CO_2 emissions	Product quality, logistics costs
Zhang et al. (2021)	SCM	Long/short lifetime perishables	DES	Food waste[a]	Selling price, revenue

Stream: DIST—Distribution, PROD—Production, SCM—Supply chain management; *modeling and simulation approach:*ABM—Agent-based modeling, DES—Discrete event simulation, HS—Hybrid simulation, MCS—Monte Carlo simulation

[a]Indirect indicator for energy performance (e.g., can be statically converted into emissions)

6.4 Applications

To demonstrate the applicability of simulation methodologies for assessing energy-related aspects of perishables, the following section elaborates on three distinctive use cases. The first case (Sect. 6.4.1) combines a DES approach with a life-cycle assessment procedure to improve decision-making for juice production systems and quantify energy consumption as well as waste, air pollutants, and ethene emissions. Using the exemplary case of regional strawberry supply chains in Austria, Sect. 6.4.2 elaborates on the integration of food quality models and inventory management policies to allow for estimating the quality of fresh food items depending on storage temperatures and durations. Finally, the third use case presents a simulation tool that is based on a discrete time approach. The simulator is leveraged to assess the media consumption of a beverage bottling plant in Germany, outlining the implications of different system designs on various media such as energy and water consumption or CO_2 emissions.

6.4.1 Juice Production in Sweden

The food production industry is one of the largest industries in the world. In 2021, its production systems were responsible for a third of global anthropogenic greenhouse gas emissions (Crippa et al. 2021) contributing to a larger part of our planet's environmental footprint (Ritchie and Roser 2020). These two factors make the reduction of the environmental impact from food production of crucial importance according to reports by Jödicke et al. (1999), reiterated by Bajželj et al. (2014). This is relevant for other environmental issues including global warming, eutrophication, and acidification. This alarming increase in global environmental footprints persuades general interest within science and technology to develop more-sophisticated tools and measurements for assessing environmental effects. Since these environmental footprints in food production industry occur before, during, and after the production processes (Ma et al. 2010), it is imperative to simultaneously assess and plan productivity in conjunction with environmental parameters. The authors' interest in this juice production case has been to propose a method and a tool for use in effective production planning with reflection to environmental parameters on the same basis and same time frame, as well as to unify the output data on the same framework. This is to prevent the high risk of sub-optimization (Calvet et al. 2020), if only parts of the relevant parameters are considered at a time. Focus on climate changes and other related environmental phenomena are growing, as its stakeholders become more aware of how production processes and utilization of resources from the earth affect the environment with noticeable efforts by international standards (ISO 2006a, b). Lind et al. (2008) released their "Green Supply Chain Modeling Solution" for carbon footprint simulation as a solution to calculate the GHG footprint, which suggested ways to reduce emissions, proposing the incorporation of carbon offset purchases into production and footprint calculations. The models obtained by Kogler and Rauch (2019), Wohlgemuth et al. (2004), and Gäbel and Tillman (2005) are previous examples of models resulting from discrete event simulation use in different areas. Although there has been a gradual progress in the development of such models, most analyses made with this technology focus on process performance as in Azapagic and Clift (1999), where production efficiency is measured on economic goals. However, little is known on DES involving production processes and their resulting environmental impacts. This is also in line with issues previously indicated by Johansson et al. (2003). Several DES strategies on how to incorporate energy-related factors when modeling the production processes focusing on process parameters with impact to system's efficiency from the energy perspective are available. For example, Kohl et al. (2014), Solding and Thollander (2006), Mani et al. (2013), Eriksson (2014), and Vicino (2015) show examples on how to create and use DES with more emphasis on energy consumption together with system improvements. The work by Johansson et al. (2008) presents a simulation methodology towards DES that accounts for both the production and environmental parameters. In this methodology, the production parameters like batch size, batch frequency, production planning, and resource management were mutually simulated with the environmental parameters such as emissions, waste, and energy consumption. This resulted in an all-in-one output as intended, with the output data unified in the same platform. During this work, evaluation of various methods and scenarios were used in lifecycle assessments as supported by reports by Banks et al. (1996), Andersson et al. (2012), and Alexander et al. (2000). Productivity was evaluated as a measure to all process parameters of material, energy, waste, and the pollutants generated simultaneously. Three case studies were conducted, one of which was used to establish a common methodology as a benchmark for sustainable food production.

Furthermore, additional steps in simulation were made to include four additional parameters, which are raw materials, machines, facilities, and transportation in the simulation process. During this activity, additional scenarios were assessed supported by Johansson et al. (2008) and Law and Kelton (1999) in simulating data inputs of environmental parameters from lifecycle assessments in combination with production data. Changes on the manufacturing floor were also analyzed to resolve simulation constraints. Results show that by performing this type of DES, some changes within the system could improve both environment and the productivity parameters, while in some cases the changes have improved one parameter while aggravating the other. The results are used as a baseline for decision-making in the juice production system. This addition also provides information on both the input and output variables of the production process, where material and energy consumption are input and waste, pollutants such as CO_2, NO_X, SO_2, and ethene are generated as output.

6.4.1.1 Juice Production Process

The production system of juice principally consists of purée machines, mixing tanks, heat exchangers, buffer tanks, packaging machines, palletizers, as well as conveyors, pumps, and an industrial piping. In this case study, the system produces about 25 million liters of juice annually consisting of an aseptic package line for fruit juice, soups, and compotes and with a system comprising up to three purée machines, two mixing tanks, three heat exchangers, six buffering-tanks, their palletizers, five packaging machines, several conveyors, pumps, as well as a huge number of pipes within the facility (Fig. 6.2). The facility also incorporates an incoming process inventory with freezers located at 20 km from the production line.

To process fruits into juice, the raw fruit is mushed into a thick liquid suspension or paste, which is generally called purée. This is conducted using the purée machine. The paste is then mixed with other ingredients for a distinctive taste. The resulting mixture is afterwards pasteurized and pumped into the buffer tanks. After the buffering process, the buffered mixture is sent into the packaging machine. To satisfy the market needs and customer demands, the processed juice is packaged in different shapes and sizes and finally loaded on pallets. From there, delivery is either made directly to customers or through warehouses as stocks to be delivered to future customers. After the dispatch of products, the leftover paste in the pipes is discharged as waste, characterized as low-value products, or recycled to the incoming process

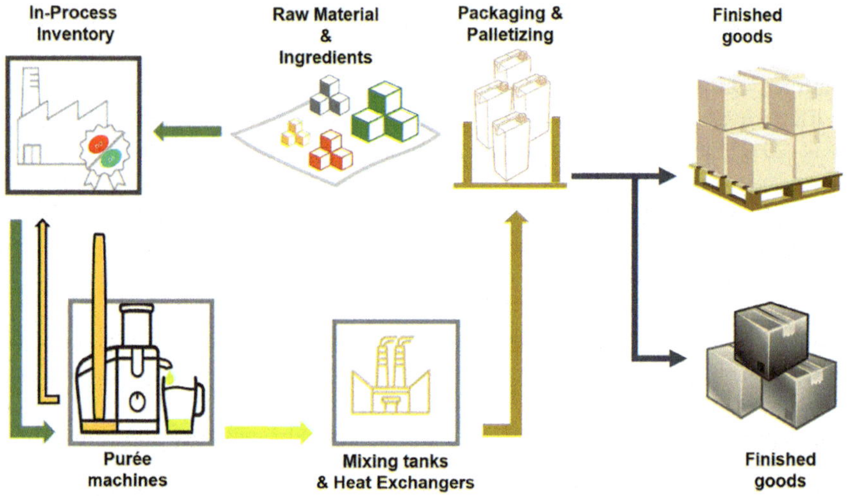

Fig. 6.2 Overall description of the production system, combined DES-LCA model of the system

inventory units to be frozen and reinjected into the production line if considered as high-value product.

6.4.1.2 Simulation Model

The production system is simulated with normal production data like flow speed in pipes, setup times of machines, shift schedule, tank volumes, machine speeds, speed on conveyors, palletizing times, etc. At each step, the model is also incorporated with environmental impact data. This type of model is referred to as the cradle-to-gate Life-Cycle-Assessment (LCA) data and used for the incoming materials and ingredients like:

- Apples
- Oil
- Oranges
- Paper
- Water
- Aluminum
- Polypropylene
- Electricity

Within this model, individual ingredients are described with respect to five features characterizing the environmental impacts, which are:

- Acidification potential measured in grams of SO_2 equivalent (g SO_2 eq.)
- Eutrophication potential measured in grams of NO_X equivalent (g NO_X eq.)

- Energy use, measured in megajoules (MJ)
- Global warming potential measured in grams of CO_2 equivalent (g CO_2 eq.)
- Photo oxidant formation potential measured in grams of ethene equivalents (g-Ethene)

To run the simulation, the ingredients and materials are entered into the model each with their collective values for environmental impact amongst the above five features of emissions. The assessment of each event is performed from the perspective of the resource rather than the product perspective: A machine wasting a part of its product releases as much environmental pollutants as if it is operated. If two different machines with the same functionality were to be evaluated, a comparison between them would be made based on their environmental effects. This is confucted by first running the simulation with one machine and then making use of only the machine in comparison. Figure 6.3 shows a snapshot of the simulation model made in Rohrer (2003) of the juice production model from Johansson et al. (2008). The tanks are represented by the structures in cylindrical shapes, while the filling and packaging machines are represented by the large square objects and the conveyors correspond to the line between them for loaded packages. The pipes are not shown in Fig. 6.3. In the lower part of the picture there is a large conveyor that marks the exit point of the system, where the juice packages loaded upon euro pallets exit the production system during delivery.

Fig. 6.3 Screenshot of the simulation model

6.4.1.3 Results and Discussions

Several analyses have been conducted and some additional ones are in progress. The most interesting results are briefly described and discussed. To ensure that the model's reality matched sufficiently the real-world system, a reference scenario has been set as a baseline with data from the past years' production schedule. To validate the model, a comparison of one year of simulated output data from the production orders was run. This output was compared to the actual output from the factory according to Sargent (2000). The initial test was used as a validation model with settings of the production year 2006. Comparisons between the models have been performed with respect to the real-world system as presented in Table 6.2.

The variation between the model and the real-world data is acceptable, as the model excludes other products (wine produced in the real factory is excluded and constitutes 12% of the production) in the analysis. Critical faults resulting from excluded machines in the analysis justify the low percentage of waste in the model, which is equivalent to 0.3% of waste considered in the real-world system. During discussions with personnel at the production company, the model was judged as valid, on the basis that the interval of confidence was satisfactory when considering the excluded products not considered in the analysis. When the base model is run in continuous mode (24/7), the standard deviations between consecutive runs were less than 2%.

After a valid model was achieved, the model was used to simulate the system in continuous mode to compare it with alternative solutions on how to operate a juice production system. Several tests made with variations, for instance, more or fewer tanks, smaller or bigger batches, less resetting, additional resources, less breakdowns, etc. A detailed discussion is published by Persson and Karlsson (2007). A selection of interesting results is presented here. The company was most interested in a comparative analysis to outline the necessary changes of batch sizes, as this could result in the reduction of production times. Figure 6.4 describes the relationship between the batch size and the production time. From this figure, lowering the batch size shortens the production time. However, to determine the total number of hours necessary for producing products annually, it is important to obtain a more-precise or minimum batch size.

Figure 6.5 presents the relation between batch size and product waste. An increase in waste results when the order size becomes too low. The figure also shows that there is no significant change when larger batch sizes than 20 are produced. However,

Table 6.2 Model validation

	Simulation model data indexed versus real world (%)	Real-world data (%)
Raw material (kg)	88	100
Finished goods (kg)	88	100
Waste difference	0.3	

Fig. 6.4 Manned production time versus batch size

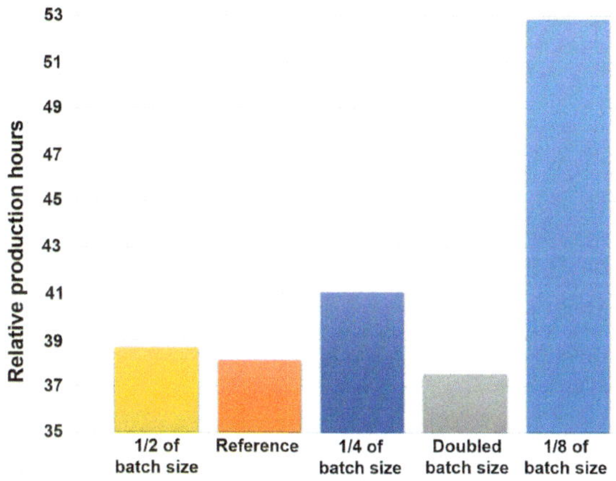

lower batch sizes increase the waste significantly, because of the products lost due to the cleaning of pipes and tanks. In Fig. 6.5, each dot represents one simulation experiment with several runs.

To optimize the production, the next step has been to determine the machines or production parts limiting the production process with attention to the most critical ones, following the theory of constraints described by Goldratt (1990) and decision support tools by Kheawhom and Hirao (2004). To realize this, the resource utilization from the reference model was considered. By adding either one or more parallel resources to the system at possible bottleneck locations, different solution options have been examined. Prior to the simulation, it had been envisaged that it could be reasonable to add another mixing tank and an additional operator. However, the simulation results revealed that this measure would not yield a positive result (Fig. 6.6).

Fig. 6.5 Percentage waste of batch versus indexed batch size

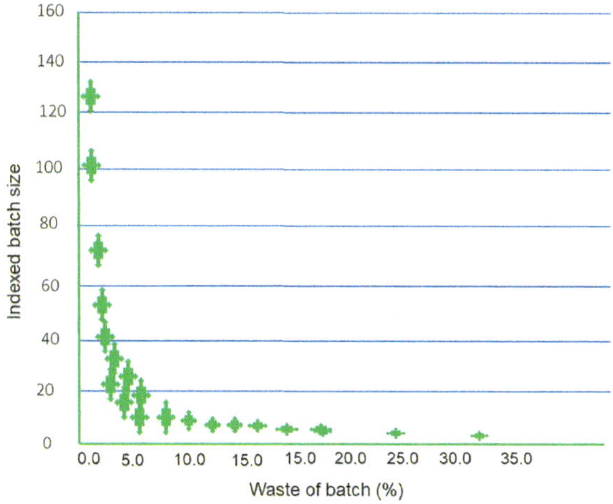

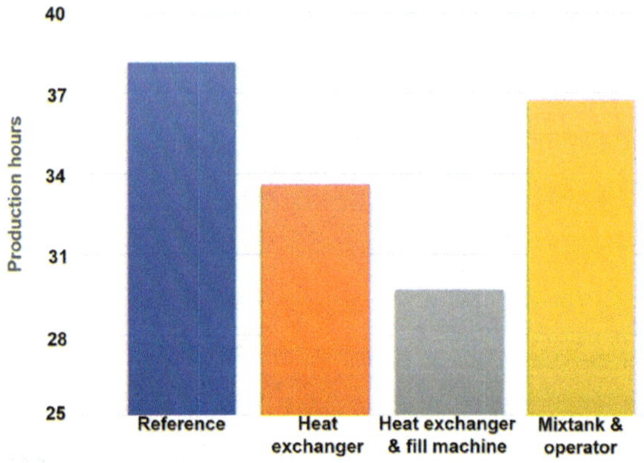

Fig. 6.6 Production hours per resource

The addition of either a heat exchanger or a heat exchanger together with a new fill machine yielded more interesting results.

The previous considerations have followed the usual intensions to study a typical DES. In the next step, the combination of a typical DES analysis with that of the environmental analysis from the LCA perspective is put into focus. To simultaneously decrease the environmental impact while reducing the lead times in production, more emphasis has been put on adjusting the batch sizes and their sequence. The resulting environmental footprint related to the production system with a specific batch size and sequence could be accounted for an annual production as shown in Fig. 6.7. In this figure, the reference refers to the current batch size and sequence adopted in the production system. In case A, the batch size was doubled for the same sequence compared to case B, where the batch size was halved again for the same sequencing, whereas case C represents the minimal or theoretical global warming potential for optimizing both the batch size and sequence. For direct application, case C would be unrealistic, as it leads to considerable stock of finished goods, although it helps to set a target for achieving low global warming potential and reduced production time.

Furthermore, an analysis has been conducted using this model to determine to what extend the in-process inventory was unfriendly to the environment. This was achieved because of the comparison of outcomes by either (1) keeping the current system unchanged (with in-process inventory located 20 km away for the main production facility) or by (2) skipping the reuse of in-process inventory, discarding or wasting the additional purée sequentially prior to the completion of each batch. The results of this in-process analysis show that it is preferable to reuse the additional purée within the system—which involves freezing and shipping processes—as opposed to wasting the purée, which consequently leads to considerable environmental effects (in a ratio of 1 to 400 in additional global warming potential) in the production system.

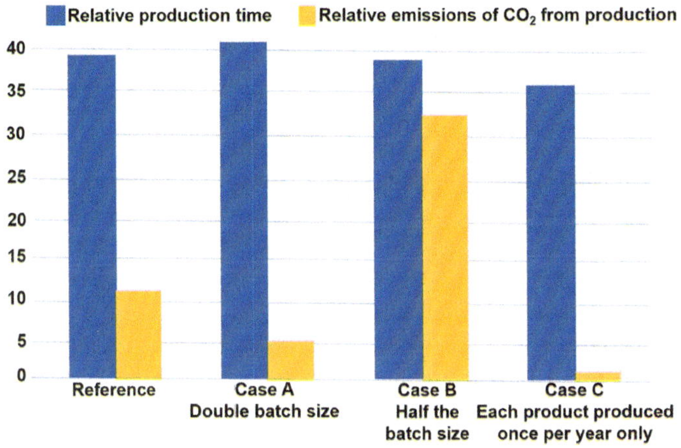

Fig. 6.7 Relative production in relation to global warming potential

6.4.1.4 Lessons Learn

Steps have been taken to analyze complex production systems with emphasis on the effects of production systems in the environment. A simulation experiment tool was developed that helps to optimize the complex processes. The use of this tool ensures analysis and optimization of both process and environmental parameters of the production system, and acts as a solution to suggest needs for improvements when diagnosing existing process lines. In comparison to previous methods that have considered the analyses of environmental and process parameters as separate entities, this tool has proven to be more time-efficient. The authors also find this method more dependable, since all results are obtained in the same unified data framework. Therefore, due to today's production‘s impact on our environment, environmental considerations should be rated as non-dissociable parameters when planning, designing, or evaluating a company‘s overall performance.

6.4.2 Food Waste in Regional Strawberry Supply Chains of Austria

The logistics of fresh fruits and vegetables (FFVs) is challenged by various uncertainties present in supply, demand, and product characteristics (Fredriksson and Liljestrand 2015). Consequently, simulation models are frequently used to provide decision support and derive managerial implications on how to guarantee food security, reduce costs, and act more environmentally friendly. An overview on related work can be found in Soto-Silva et al. (2016), Utomo et al. (2018), and Fikar (2020).

Of particular interest in this context is the challenge of food losses and waste, whereas the former refers to the quantitative and qualitative loss of food from harvest up to—but not including—retail and the latter to food wasted at the retailer and consumer household levels (United Nations Environment Programme 2021). According to recent studies, 14% of all food produced is lost (FAO 2019) and 17% wasted (United Nations Environment Programme 2021), highlighting the great need for innovative tools to facilitate more-sustainable operations in the future. For FFVs, this is of particular importance as loss levels are generally higher for such food types compared to other categories (FAO 2019). The following part provides a use case on how simulation modeling can facilitate better decision making in food supply chains and, consequently, how it can contribute to achieving the goal of reducing food waste levels and related energy consumption.

6.4.2.1 Regional Strawberry Supply Chains

As a sample setting, a regional supply chain for organic strawberries as introduced in Leithner and Fikar (2019) is considered in this work. Berries are highly perishable products with short shelf life and, consequently, require special attention during food logistics processes (Nunes et al. 2014). In the investigated setting, the products are harvested at multiple farms throughout the study region and are subsequently transported to cold stores for initial cooling. At such cold stores, the quality is checked. If a certain quality threshold is passed, the items are shipped from the cold store to a central distribution center, where the quality is once again checked to see if it matches the specified requirements of the retailers. Next, strawberries are shipped to various retail locations, where customers can buy the products for consumption. If no purchase occurs, the items are considered as food waste. An overview of this setting is provided in Fig. 6.8.

In this work, the focus is set on benefits of providing cooling and its impact on lowering food losses and waste. Ideally, the strawberries are cooled immediately after harvest, and a low temperature is maintained throughout the entire product life cycle. In reality, however, strawberries are often subject to higher temperature at various

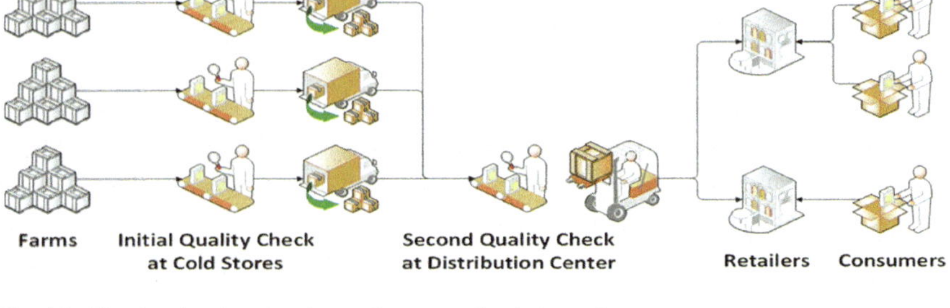

Fig. 6.8 The simulated regional strawberry supply chain setting

points in the supply chain. For instance, harvested items are often waiting on the field until the vehicle is full, and cold stores are not always available in close proximity to the harvest location. This lack of pre-cooling time is highly critical to the remaining shelf life of berries, as the products often experience high quality losses in this stage (Nunes et al. 2014). By developing a simulation model, the impact of such a common lack of cooling facilities in FFVs supply chains can be investigated.

6.4.2.2 Modeling and Simulation Approach

The problem is investigated by the development of a simulation-based decision support system, expanding on the work presented in Leithner and Fikar (2019). It integrates food quality models and inventory management policies within a DES to allow for estimating the quality of fresh food items depending on storage temperatures and durations. At the center of the work stands the user, who defines various input parameters concerning the supply chain setting, demand data, available resources, and product characteristics. This forms the input to the simulation, which models multiple weeks of strawberry supply chain operations. At the end of the experiments, the user is provided with various key performance indicators stating, among others, generated sales, service levels as well as resulting food waste levels. Figure 6.9 gives an overview of the employed modeling framework.

Within the simulation, each strawberry is modeled as an entity with an individual initial quality and parameters defining the spoilage rate of the item at various temperatures. Consequently, some items may be more susceptible to high temperatures than others. Throughout the product life cycle, the storage durations within the individual steps of the supply chain are recorded and the expected current quality of the items is constantly updated. If it falls below a certain threshold, the items may be removed

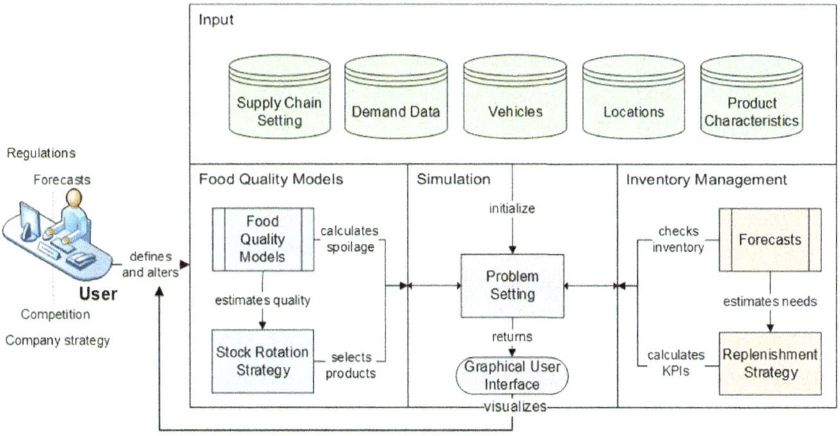

Fig. 6.9 Decision support system to reduce food waste in regional food supply chains

from the process at one of the previously defined quality checks. Consequently, four different end situations may occur for each strawberry: (i) it is bought by a consumer, (ii) it is removed at the cold store level for alternative usage; (iii) it is removed at the distribution center for alternative usage, or (iv) it is wasted at the retailer. Food waste at the consumer level is not considered in this work.

Queues are used to model inventories. To decide the order of the queue, i.e., which strawberries are shipped first to the subsequent supply chain stage, three common stock rotation schemes are implemented. The first one, first expired—first out (FEFO), focuses on reducing food losses by always shipping items with the shortest remaining shelf life first. Last expired—first out (LEFO), in contrast, prioritizes items with long remaining shelf lives to offer the customer high-quality products. Additionally, the decision on which items to ship first may be randomized without following a specific decision rule.

All movements within the simulation model are performed by resources with a specified capacity. While items are shipped from the farm to the cold store and to the distribution center based on a predefined shipping schedule, shipments to the retail stores are based on orders following a base stock policy. Once the items reach the retail stores, they can be bought by a customer. As the customers' picking behavior has a major influence on food waste (Teller et al. 2018), the three previously introduced picking settings are again integrated within the simulation framework to model which items are selected first by the customers. At this point, the strawberry process ends, i.e., any additional steps after the customer's purchase at the retailer are not considered.

6.4.2.3 Results

The following computational experiments investigate the impact of having cooling equipment available at various stages in the supply chain on food waste. The base setting and all input data are based on the work presented in Leithner and Fikar (2019). It consists of 59 organic strawberry farmers, 359 retail store locations and a single distribution center, all located either in Lower Austria or Vienna. The model was developed with the simulation software AnyLogic 8.7.9 with all procedures coded in Java. Reported results represent the average value over 100 replications for each investigated setting, corresponding to around 248,000 simulated strawberries per run. Each replication starts with a warm-up phase consisting of eight weeks of operations to set up the system. Only strawberries harvested in the subsequent weeks 9 to 12 are counted to the statistics. The simulation stops once all strawberries harvested during these weeks reach their final state, i.e., have left the process.

Within the first set of experiments, the number of cold stores is varied from having only a single cold store for all farmers in the region to having a cold store at each farm location. The former setting results in lower fixed costs, however, substantially increases transport and further leads to additional food quality losses due to the lack of pre-cooling compared to the latter one. In case of a limited number of cold stores, the farms running a cold store are set randomly. Farms without a cold store ship all strawberries to the nearest one. No capacity constraints at the cold store are set in order to enable this experiment and alternative distribution channels are not considered in order to solely focus on the impact of pre-cooling. All three demand priority settings of the customers (FEFO, LEFO, random) are investigated with FEFO implemented as the stock rotation scheme of the distributor.

Figure 6.10 shows the results of these computational experiments with each point representing the average percentage of reference products wasted over all replications. As the results highlight, the customers' picking behavior in the store greatly impacts food waste. If customers always take the items with the longest shelf life first, i.e., follow a LEFO strategy, food waste levels are substantially increased compared to a random or FEFO picking behavior. This factor, however, is out of the control of a regional food supply chain. Consequently, companies need to consider different ways to reduce food waste levels.

One potential option is to invest in additional cooling equipment. Therefore, the second set of experiments investigates the impact of having additional cooling equipment available at retail level. Strawberries are commonly kept in stores at ambient temperatures within the study region (Leithner and Fikar 2019), substantially reducing the keeping quality of strawberries over time. Within the first set of experiments, a storage temperature of 20 °C at retail stores was assumed. Figure 6.11 shows the impact on food waste of reducing this storage temperature to 3 °C in increments of one. Each dot represents again the average value of 100 replications for each setting.

As the results show, the storage temperature at the retailer has a large impact on food waste levels, again highlighting the importance of having proper cooling equipment available in FFVs supply chains. If the customers follow a LEFO strategy, food waste reaches the maximum, followed by random and FEFO. Additionally, one can derive from this figure that the reduction resulting from a decrease of 1 °C in storage temperature changes based on both the initial temperature and the customer picking behavior in the store. Additionally, it indicates a common trade-off present in perishable food operations. To reduce food waste, additional cooling is required, which leads to an increase in energy consumption. Simulation models enable to

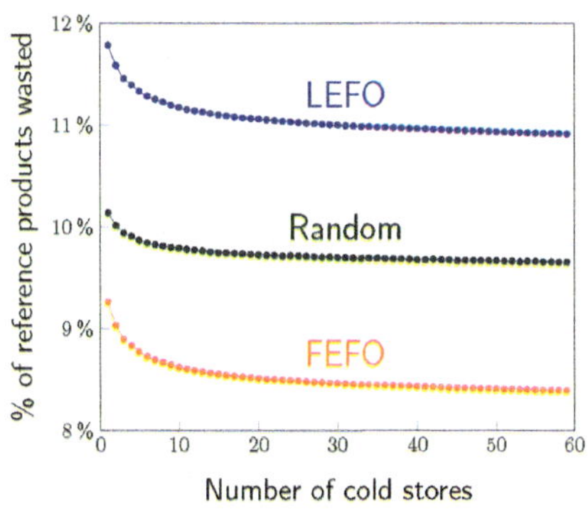

Fig. 6.10 Impact of having pre-cooling available at farm level on retailers' food waste levels considering various customers' picking behaviors in the stores

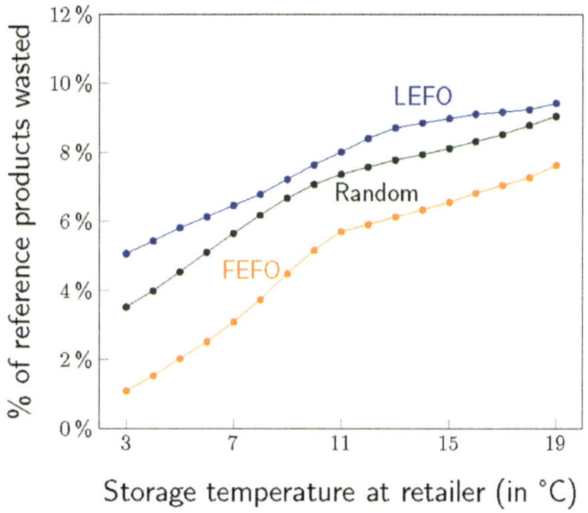

Fig. 6.11 Impact of having cooling equipment available at retail level considering various customers' picking behaviors in the stores

analyze such trade-offs in detail to improve decision making and act more sustainable in the future.

6.4.2.4 Lessons Learn

As the results of the computational experiments show, food waste in regional food supply chains is impacted by a wide range of influencing factors. This includes the available infrastructure as well as the implemented decision rules and how customers pick items in the retail store. Relating to the specific use case introduced in this part of the work, the great need to provide efficient cooling equipment throughout each step of the supply chain is once again highlighted. Consequently, both policy and decision makers need to understand the complexity of food operations and how the individual influencing factors relate to each other. Simulation-based decision support systems enable such a holistic understanding and further allow for testing new approaches in a flexible and risk-free manner. This further provides various options for future work. For instance, this work could be expanded by incorporating customer behavior and adjustments over time through an ABM approach or integrate multiple types of FFVs simultaneously to study how various product mixes and assortment strategies influence food waste in retail settings. Additionally, investigating the trade-off between the additional energy consumption required for cooling and food waste is of interest.

6.4.3 *Energy and Media Consumption of a Bottling Plant in Germany*

The food and beverage industry is caught between the conflicting priorities of high product quality, steadily increasing product diversity, and cost-effective production. Increasing energy efficiency and more-flexible production routines can sustainably reduce costs and, thus, increase competitiveness. Due to the foreseeable further increase of energy costs, the current and future ecological challenges, the associated growing public awareness, as well as legal developments, operators of manufacturing and packaging plants are faced with the complex challenge of ensuring energy-efficient and sustainable production routines. On this background, particularly cost-driven businesses in the beverage and food industry have begun to analyze their own energy and media consumption and look for optimization strategies in this area. Energy costs already rank third in the operational cost structure in the food and beverage industry. Therefore, monetary considerations support this development, too. However, measures to effectively reduce energy and media consumption have often only been taken locally for individual systems. In this context, the discussion of sustainable production must take into account various aspects, which need to be considered simultaneously. In addition to the material efficiency of the substances and means used in primary production, energy efficiency plays an increasingly important role. Material efficiency requires knowledge about the current production processes. Simulation, which can be used as a standardized tool for optimizing processing and packaging plants, can also be employed to improve energy efficiency.

6.4.3.1 Beverage Bottling Plant

The given use case outlines a method for the automatic generation of simulation models for the holistic simulation of energy and media consumption for beverage bottling plants. Figure 6.12 synopsizes the main steps of the use case approach. Using a comprehensive and complex database with a suitable underlying database structure, a beverage bottling plant could be modeled and parameterized in terms of physical plant, articles, and production schedule. The simulation model is automatically generated via an XML-based configuration file in a discrete time simulation environment. Finally, a holistic production plant can be simulated considering all energy and media requirements including a production plan. Thus, with the inclusion and consideration of a production schedule, a description of the energy consumption during inactive production times can also be made.

The object of this use case is an industrial glass refillable beverage bottling plant for mineral water and soft drinks of one of the ten largest mineral water companies in Germany (in terms of sales). The main product of the plant is mineral water. The plant is schematically outlined in Fig. 6.13. It is divided into three sections (material flows), depending on the material being transported (pallets, crates, and containers or bottles). The interlinking of conveyors allows for buffering downtimes that are

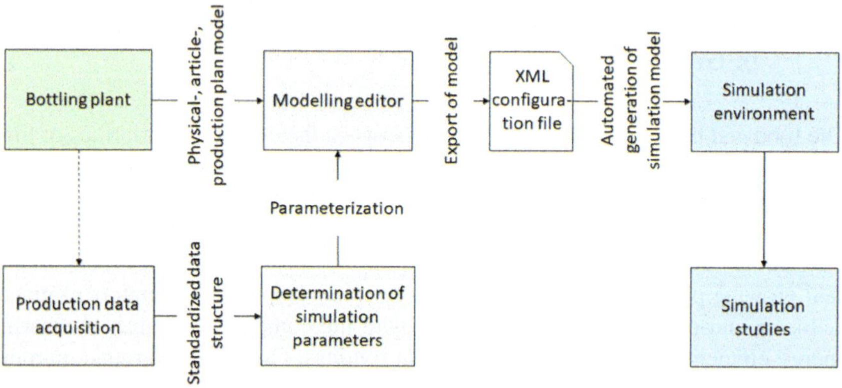

Fig. 6.12 Overview of the main steps of the bottling factory case

caused, for example, by the failure of individual machines. If the buffer capacities of the conveyors are exceeded, the respective malfunctions and downtimes lead to a fault propagation, which is characterized as a shortage or backlog situation.

The so-called empties, which are bottles in crates, are fed to the line on pallets, and the crates are unstacked by the depalletizer. The crates with the empty bottles pass through a so-called cap unscrewer, before the glass bottles are unpacked by an unpacking machine onto the conveyor for containers and bottles). Then, the bottles are transported to a bottle-cleaning machine where they are cleaned and disinfected. Subsequently, they are inspected by an inspector machine for damage and remaining contaminants, and then filled with liquid and capped in the filling and capping machine. The filled bottles are inspected in the full bottle inspector for glass fragments and other contaminants, before they get labeled in a labeling machine. In the further process, the bottles are packed into the previously cleaned crates in the crate filler and stacked on pallets by the palletizer. The material flow of the empty crates is abstracted in the model and, thus, represented in simplified form by a crate washing machine and a crate magazine. which serves as a buffer system. In the real plant, a complex throughput system for crate transport is employed at this point. In the material flow of the pallets, the pallets are also buffered in a magazine.

The plant is connected to an in-house data acquisition system (DAQ). Every two seconds, the operating mode, operating status, and power consumption of each machine are recorded. In addition, thermal energy, water, CO_2, and compressed air requirements are recorded for the entire plant. The data points follow the declaration of the Weihenstephan standard, which is a standardized data interface between the individual machines and a higher-level system (Kather and Voigt 2005). It defines the unique and standardized semantics of the individual data points as well as the transmission procedure. This makes it possible to receive and process equivalent data from any machine, regardless of the manufacturer. The data sets of the returnable glass bottling line used in the application case comprise two measurement campaigns, the first covering a period of two months with the above data scope. In contrast to the

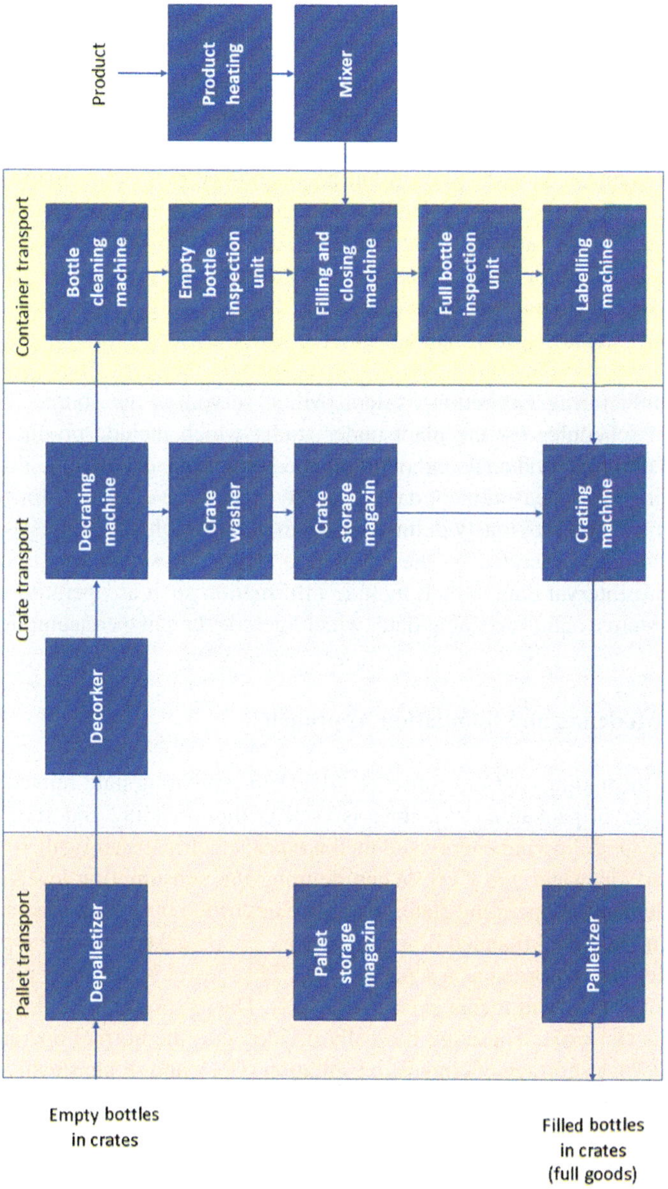

Fig. 6.13 Schematic representation of the beverage filling line under study

period of the second measurement campaign, the first data collection was characterized by almost continuous production, apart from the weekend. In the second data collection period covering four weeks, the beverage filling plant was only in operation on approximately three days per week. The individual units were additionally equipped with measuring devices for recording the compressed air requirement and these were connected to the in-house data acquisition system. Other machines such as the empty bottle inspector, the full bottle inspector, and the unscrewing machine were also included in the ongoing measurement. The only consumer of thermal energy and water is the bottle washer. Its warm wastewater is used for the crate washer. Thus, the latter machine does not require any thermal energy on its own. The filling machine, in turn, is the only machine that requires CO_2 for the production of certain products.

A Manufacturing Execution System (MES) served as the source for detailed production schedules for the plant under study, which include production times, quantities of items produced, and other planned downtime due to, e.g., maintenance and cleaning. The measurement data from the DAQ were converted for the project into a uniform and previously defined data structure, which also serves as the structure for the data generated by the simulation. The data within the structure were divided into interval data, which include information such as operating mode and operating status, and time stamp data, which include data like consumption values.

6.4.3.2 Modeling and Simulation Approach

Following the findings of Osterroth et al. (2017), the operating state-related consumption behavior of packaging machines is used in the modeling and simulation and assigned to the following energy and media types: electrical energy, thermal energy, compressed air, water and CO_2. When defining the consumption levels, the operating mode and the operating state, which are described by a state model, are used. In addition, so-called planned downtimes, e.g., on a weekend when no production is scheduled, are recorded as a separate state. The reason is that most plants have a demand for energy and media even in this case. During production, different operating states can exist. These are basically divided into the normal operation of the machine (C_L), a shortage or congestion situation (D_{L1}), and stochastically occurring failure situations (D_{L2}).

For an automatic determination of simulation parameters, such as consumption quantities as well as parameters for the description of the failure behavior, a suitable evaluation software was developed. The parameters can be defined specifically for different evaluation criteria, such as article-based, container-type-based (bottle format) or time-period-based. The average consumptions of the individual energy and media types are calculated in the intervals in which the operating state does not change. In addition to the machine-specific determination of the total consumptions, a detailed statistical evaluation of the consumption values (weighted average, standard deviation, variance, standard error, and 95% confidence interval) was carried out. The failure behavior is determined on the basis of the operating condition of a machine. The *time to repair* (TTR) denotes the duration of the failure, while the failure-free runtime between two consecutive failures is defined as *time between failures* (TBF). To determine the mean TTR (MTTR), the times of D_{L2} states are used for calculation, and for the mean TBF (MTBF), D_{L1} as well as C_L states are combined. The availability results are shown in Eq. 6.1.

$$\text{Availability} = \text{MTBF}/(\text{MTBF} + \text{MTTR}) \tag{6.1}$$

In addition, the operating times and the downtimes are investigated with regard to the distribution curve of the failure behavior (failure distribution behavior). The suitability of the Weibull, negative exponential, and lognormal distributions was investigated by Voigt and others (Voigt 2004), with the Weibull and exponential distributions achieving the best fit. In this application, the negative exponential distribution is used, because random failures are assumed with a constant failure rate and this describes the behavior sufficiently accurately. Moreover, the exponential distribution can only be described by a single variable (β). In addition to the data from the DAQ, it is necessary to determine other parameters directly in the plant. Therefore, the topological data of the plants as well as of all machines and means of transport (conveyors) were determined within the scope of performance analyses according to DIN 8743. The set output of all individual aggregates was analyzed manually over the data acquisition period, e.g., by the aid of measuring light barriers. Furthermore, the outputs per filling article of the individual aggregates and the global reject rates of inspection machines were determined. All buffer lines were examined with regard to their buffer capacities and times.

The modeling of processes and material flows in the form of a simulation model represents major obstacles for many small and medium-sized enterprises (SME). Therefore, a modeling editor by Bär et al. (2021) was developed and used within the project. During the development of the editor, special attention was paid to a user-friendly and simple application in order to enable modeling even for users without prior qualifications in the field of simulation. The editor generates a standardized XML-based configuration file for a subsequent simulation environment, which contains all required simulation-relevant parameters and information about the model structure. The parameters have to be entered manually in the modeling editor. Numerous auxiliary functions support the user, which simplify and accelerate the modeling process. The modeling approach implements existing standardizations, such as ANSI/ISA S88. The models consists of a model for the physical representation and illustration of plants, a process model for the representation of batch production processes by an article model, and a production plan model. The modeling depth of the editor goes down to the machine level (e.g., filling machine) and is mainly related to the scope of the required parameters as well as the complexity and accuracy of the models. The article model enables the modeling of different filling articles, which can be produced on the so-called process cells of the factory model. Each machine can be parameterized in terms of its parameters describing consumption and performance, but also in terms of the type of container as well as the packaging ratios.

The production plan model describes the production to be processed in the simulation. It includes a shift schedule to define production and non-production times, such as a weekend when no production is planned. Furthermore, a sequence plan is defined, which determines the sequence of items and quantities to be produced. In a setup matrix, the required changeover and cleaning times that are needed between the articles of the sequence model are defined.

All simulation studies were carried out using the “PacSi” simulation environment, which was developed by SimPlan AG, Dresden, Germany, as an industry-specific simulation system for processing and packaging plants. The software uses a discrete time approach as the simulation method. The model creation within the software is based on building blocks that differ fundamentally in functionality, separation, and merging of material flows and parameter sets. Stochastic failure behavior is assigned to the individual elements and failure propagation is achieved by chaining the machines directly or via conveyors between machines. The failure behavior is described by varying the seed value using a random number generator that initializes a random number stream. Thus, these streams for MTTR and MTBF, among others, are regenerated according to the distribution function (negative exponential) at each simulation run and the times and durations of the failures are varied. Statistical reliability can be achieved on the one hand by long simulation durations and on the other hand by multiple repetitions of the experiments.

In contrast to other approaches, the use of a special industry-specific simulation system features the economic and quality-assuring advantage that proven, repetitive procedures are effectively automated and prefabricated. Moreover, validated partial models are available for the representation of reality. The smallest partial model that can be mapped to a simulation system is called an element. These pre-programmed elements (e.g., machines, conveyors, storage) are archived in the form of a library and made available for the elementary configuration of structures of the processing plant.

In the context of this use case, the simulation environment has been extended by numerous functions. The most important extension is the automatic generation of simulation models based on the XML configuration file of the modeling editor. For this purpose, the various specific building blocks of the modeling editor are mapped to the existing building blocks of the simulation environment. By using a coordinate system as well as the edges of the elements in the modeling editor, the topology and logical structure can be interpreted in the simulation environment. When the simulation model is created, the parameter sets of all units are loaded from the configuration file and parametrize the simulation model. A special feature is the integration of a production schedule, comprising the shift schedule, the sequence of different items, and the changeover matrix. This is mainly solved by sources, i.e., elements in which material is produced in the simulation, by storing the non-production times as well as recording the times between articles and their specific article number. An indexing allows an element-specific parameter change and enables an article-specific simulation within the production plan. Another special feature is the definition of non-production times by the shift model, as the newly implemented off-consumption level (OL) is active during this time. The consumptions in the production times are described by the respective operating state. The simulation results generated every second are stored in an SQLite database.

In the PacSi simulation system, a parameter mask (Fig. 6.14) can be used to specify the respective consumption values (Table 6.3), determined from the consumption measurements for each consumption medium for the defined operating states. In addition to the constant basic consumption, a linear consumption behavior can be defined for each consumer for the state of normal operation (C_L) of the machine. This takes into account that machines often do not operate continuously at the respective set output due to interlinking influences (e.g., reduced feed capacity) or speed controls depending on upstream and downstream buffers. This means that the fluctuating output of the machine during normal operation can also be taken into account in the energy and media consumption simulation in such a way that a linear dependence of the consumption in relation to the output is simulated close to reality.

6.4.3.3 Results

Two different time periods are available for the validation of this use case. Validation Period 1 covers the electrical energy demand of the aggregates, simulating an extensive production schedule with production-free periods such as weekends. Validation Period 2 includes the full energy and media requirements of the beverage bottling line's aggregates. Here, a shorter period with only two items is investigated. The values describing the failure behavior of the machines were determined specifically for the corresponding period. The real data recorded by the measurement data acquisition system of the production schedule were converted into the data structure described above, and the simulation-relevant parameters were determined automatically by running the evaluation software programmed in the project. This resulted in a total of approximately 1,250 parameters for all five articles. The correct operating status messages of the individual machines were randomly checked during operation. The changeover times of the plant between the individual articles due to any necessary rebuilds or cleaning were taken from the real data during the validation periods.

The beverage bottling plant investigated in this use case was modeled in its entirety with all material flows and parameters in the editor described (Fig. 6.15). The main material flow (thick arrows) starts at the main source at the top left and runs clockwise to the sink at the bottom left. The two secondary flows (thin arrows) for pallets and boxes are on the left and in the middle, respectively. The items produced in the validation periods were each modeled with item-specific parameters.

The present model also includes article-specific and container-specific parameters. The changeover times were transferred to a setup matrix in order to define the downtimes between the individual articles. The shift times for the validation period could be determined from the real data (non-production from Saturday 08:30 to Monday 07:00) and were also parameterized in the modeling editor.

In Period 1, the focus of the investigation was on electrical energy consumption and the complex production schedule. For this purpose, the data for a production

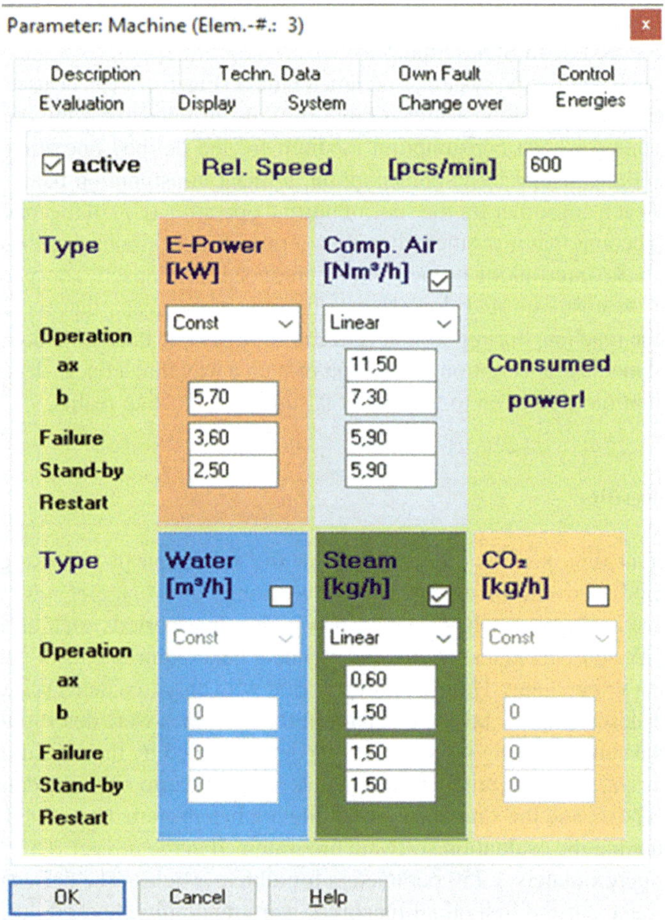

Fig. 6.14 Input mask for the parameters for simulating the energy and media consumption for the filling line (screenshot)

period covering four articles including a production-free weekend included in the recording period were recorded. Here, the filling and packaging of the individual articles differed in their total electricity requirements. The production quantities of all articles in the simulation show a deviation of less than 0.1% compared to the real production counter of the filling machine. The production times of Article 1 and Article 2 in the simulation are in a range of about 6–8% compared to the determined real times. Article 3 shows a deviation of approximately 4.5%. Article 4 shows a total deviation of 11.2%, i.e., accelerated production within the simulation. The bottle washer, crate conveyors, and mixer could be clearly identified as the main consumers of electrical energy in the simulation. They accounted for about 70% of the total electrical energy consumed and consistently showed a very small percentage

Table 6.3 Global data parameters for the simulation of energy and media consumption

Active Rel. speed [pcs/min]	Active	Function	ax	b	Failure	Standby
E-Power [kW]	☑	Constant ▼	0.0	5.7	3.6	2.5
Comp. Air [Nm^3/h]	☑	Linear ▼	11.5	7.3	5.9	5.9
Water [m^3/h]	☐	Constant ▼	0.0	0.0	0.0	0.0
Steam [kg/h]	☑	Linear ▼	0.6	1.5	1.5	0.0
CO_2 [kg/h]	☐	Constant ▼	0.0	0.0	0.0	0.0

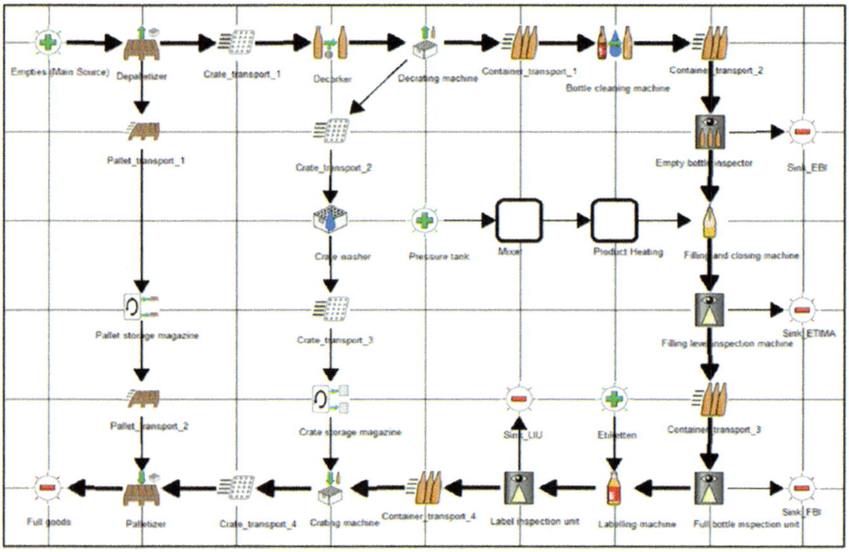

Fig. 6.15 Screenshot of the model in the editor

deviation from the measured data. The bottom 50% of the units accounted for only 12.3% of the total electrical energy. Over the entire period, a percentage deviation between the measured data and the simulation of 0.33% could be determined for the electrical energy demand of the entire plant. During the article-specific validation, it was noticed that aggregates such as the filling machine, the pallet conveyor, and the depalletizer individually showed relatively high deviations compared to the measured data. Larger deviations in the simulation compared to the measured data, particularly for Article 4 as shown above, are due to a long downtime at the beginning of the filling of the article, which can be explained by the fault messages in the measured data. Since long downtimes occur only very rarely due to the negative exponential disturbance behavior distribution and were obviously not generated in the simulation run by chance in the measured length, the simulated production runs faster and reaches the planned production quantity earlier. The simulated consumption of CO_2 within Period 1 showed a deviation of about +5% compared to the measured value. The water consumption of the bottle washer, which is the only machine with water consumption in the plant, showed an overall deviation of −11% compared to the measured values.

Period 2 was primarily used to validate the compressed air and thermal energy consumption. The simulation parameters determined here were also used in Period 1. The production quantities of the two articles filled in this period (Article 4 and Article 5) show a deviation of less than 0.1% compared to the real production counter. The production times of the two articles have a small deviation of about −3 to −3.5% compared to the production times required in reality. For the electrical energy consumption in Period 2, an overall good agreement was achieved with a percentage deviation of only 0.14% compared to the measurement results, since the bottle washing machine, mixer, and crate conveyor, which consume by far the largest part in this recording period, show only a small deviation of about 3%. Devices such as the depalletizer and the palletizer as well as the container conveyor show the largest deviations. The labeler, crate packer, filler, and empty bottle inspector, which are the four largest consumers of compressed air, account for about 70% of the total demand. The small deviations between the simulated and measured data of these devices lead to a small overall deviation of the plant's total demand (0.06%). There are larger deviations for units such as the decapper, the cycle-based depalletizers and palletizers, and the product heater. Due to the small share of the total consumption of these aggregates, these deviations are not decisive for a good estimation of the total consumption. A comparison of the CO_2 consumption values is not meaningful in this period, since CO_2 was only required for one article. The values of the heat energy demand deviate overall by +13.61% and the values of the water consumption by +6.14%. These relatively large deviations mainly result from the relatively short validation period.

6.4.3.4 Lessons Learn

An approach to modeling, automatic simulation model generation, and simulation workflows were presented to provide quick and easy forecasts of energy and media requirements for beverage bottling plants, especially for SMEs. This can be an opportunity for SMEs to achieve more-sustainable production, made possible by a modeling approach implemented in a user-friendly editor. In combination with a standardized data structure and, thus, data format, and associated evaluation software, simulation-relevant parameters such as consumption and failure behavior data can be determined automatically. However, this requires a sufficiently large database for statistical significance. The approach selected and described in the use case was applied to a concrete beverage bottling plant with a production sequence of different articles, for which a detailed production plan was additionally modeled. Based on an XML-based configuration file, the simulation model was generated automatically, which again means a recognizable time saving compared to manual model generation.

6.5 Conclusions

In this chapter, the use of computer simulation has been elaborated for the analysis and evaluation of energy aspects that are related to systems dealing with perishable items. The production, storage, processing, and distribution of perishables is characterized by a high degree of complexity, since these goods feature unique temporal, operational, and spatial constraints such as limited shelf life or special cooling requirements. Moreover, they are influenced by uncertain environmental conditions (e.g., weather, natural disasters) and strong fluctuations in demand (e.g., due to seasonality), which additionally impair planning and forecasting within this domain. Hence, to study perishables and their operational systems, scholars and practitioners require flexible, adaptive, sophisticated, and dynamic methodologies that are able to capture such interdependent relationships in a holistic manner. Computer simulation can effectively be employed to design, analyze and evaluate systems that are related to perishable items.

To position the simulation methodology as eligible and efficient assessment instrument for energy aspects of perishables, first the status quo and background of energy-related simulation research have been discussed, demonstrating its growing popularity and usefulness in scientific research. Subsequently, common performance indicators in simulation-based perishables research have been elaborated to provide guidance for future research in related industry sectors. Finally, the potential scope and applicability of energy simulation modeling was outlined by means of three real-world use cases. The first case evaluates energy consumption as well as waste, air pollutant, and ethene emission outputs of a juice production system in Sweden, while the second case outlines the use of simulation to predict food loss and waste levels in regional strawberry supply chains in Austria. The third case employs an industry-specific simulation tool to assess the media consumption of a beverage bottling plant in Germany. Based on different simulation techniques such as discrete time simulation and DES, these applications illustrate the innate capability of simulation modeling to reliably evaluate complex perishable systems. Additionally, they highlight the multifariousness as to which simulation methods can be employed within the perishables domain.

As presented in Sect. 6.2, simulation-based research of energy aspects related to perishable products is multilayered and diverse. Energy-related data on energy and media consumption can both be used as input data to specify operating conditions and as output data to evaluate the energy performance of a given system. Air pollutant emissions and food waste exclusively serve as output for the evaluation of the energy performance of supply chains and production systems for perishables. Thereby, the choice of modeling and simulation approaches primarily depends on the given system and problem domain. While discrete simulation techniques such as discrete time and discrete event simulation are very common when it comes to model the energy performance of production systems, to systemize inventory planning or management, or to optimize production planning and scheduling, an ABM paradigm coupled with a discrete time advancing mechanism can be particularly useful to assess perishables-related energy concerns that directly relate to customer demand or other properties that are based on collective behaviors and communication networks. Hybrid simulation approaches are very flexible and can be adapted to a multitude of different systems and problems, which may help researchers to develop tailored solutions for highly contextualized and specific questions. Finally, due to the fact that computer simulation can quickly become computationally intensive, the selection of an appropriate modeling and simulation technique also entails the evaluation of the individual research setting. In this regard, the use case in Sect. 6.4.3 has shown that a discrete time approach can both be technically efficient as well as conceptually comprehensive, thus allowing to evaluate the energy performance of large production systems in a constructive way.

References

Aiello G, La Scalia G, Micale R (2012) Simulation analysis of cold chain performance based on time–temperature data. Prod Plann Control 23(6):468–476

Akkerman R, Van Donk DP, Gaalman G (2007) Influence of capacity- and time-constrained intermediate storage in two-stage food production systems. Int J Prod Res 45(13):2955–2973

Alexander B, Barton G, Petrie J, Romagnoli J (2000) Process synthesis and optimization tools for environmental design: methodology and structure. Comput Chem Eng 24(2–7):1195–1200

Andersson J, Skoogh A, Johansson B (2012) Evaluation of methods used for life-cycle assessments in discrete event simulation. In: Laroque C, Himmelspach J, Pasupathy R, Rose O, Uhrmacher AM (eds) Proceedings of the 2012 Winter Simulation Conference. IEEE, Piscataway, NJ, pp 1–12

Auf der Landwehr M, Trott M, von Viebahn C (2021) Environmental sustainability as food for thought! Simulation-based assessment of fulfillment strategies in the e-grocery sector. In: Feng B, Pedrielli G, Peng Y, Shashaani S, Song E, Corlu CG, Lee LH, Lendermann P (eds) Proceedings of the 2021 Winter Simulation Conference. IEEE, Piscataway, NJ, pp 1–12

Azapagic A, Clift R (1999) Life cycle assessment as a tool for improving process performance: a case study on boron products. Int J Life Cycle Assess 4(3):133–142

Bajželj B, Richards KS, Allwood JM, Smith P, Dennis JS, Curmi E, Gilligan CA (2014) Importance of food-demand management for climate mitigation. Nat Clim Chang 4(10):924–929

Banks J, Carson JS, Nelson BL (1996) Discrete-event system simulation, 2nd edn. Prentice Hall, Upper Saddle River, New Jersey

Bär RM, Voigt T (2021) A metamodeling approach for the simulation of energy and media demand for the brewing industry. J Adv Manuf Process 3(2):1–20

Behdani B, Adhitya A, Lukszo Z, Srinivasan R (2012) Mitigating supply disruption for a global chemical supply chain—application of agent-based modeling. Comput Aided Chem Eng 31:1070–1074

Bottani E, Ferretti G, Montanari R, Rinaldi M (2014) Analysis and optimisation of inventory management policies for perishable food products: a simulation study. Int J Simul Process Model 9(1–2):16–32

Brailsford SC, Eldabi T, Kunc M, Mustafee N, Osorio AF (2019) Hybrid simulation modelling in operational research: a state-of-the-art review. Eur J Oper Res 278(3):721–737

Burek J, Nutter DW (2020) Environmental implications of perishables storage and retailing. Renew Sustain Energy Rev 133:1–20

Calvet L, de la Torre R, Goyal A, Marmol M, Juan AA (2020) Modern optimization and simulation methods in managerial and business economics: a review. Adm Sci 10(3):1–23

Chatzidakis SK, Athienitis A, Chatzidakis KS (2004) Computational energy analysis of an innovative isothermal chamber for testing of the special equipment used in the transport of perishable products. Int J Energy Res 28(10):899–916

Crippa M, Solazzo E, Guizzardi D, Monforti-Ferrario F, Tubiello FN, Leip AJNF (2021) Food systems are responsible for a third of global anthropogenic GHG emissions. Nat Food 2(3):198–209

Cuéllar AD, Webber ME (2010) Wasted food, wasted energy: the embedded energy in food waste in the United States. Environ Sci Technol 44(16):6464–6469

Czerniak LL, Daskin MS, Lavieri MS, Sweet BV, Erley J, Tupps MA (2021) Improving simulation optimization run time when solving for periodic review inventory policies in a pharmacy. In:

Feng B, Pedrielli G, Peng Y, Shashaani S, Song E, Corlu CG, Lee LH, Lendermann P (eds) Proceedings of the 2021 Winter Simulation Conference. IEEE, Piscataway, NJ, pp 1–12

De Keizer M, Haijema R, Bloemhof JM, Van Der Vorst JG (2015) Hybrid optimization and simulation to design a logistics network for distributing perishable products. Comput Ind Eng 88:26–38

Ding H, Benyoucef L, Xie X (2009) Stochastic multi-objective production-distribution network design using simulation-based optimization. Int J Prod Res 47(2):479–505

Dong Y, Miller SA (2021) Assessing the lifecycle greenhouse gas (GHG) emissions of perishable food products delivered by the cold chain in China. J Clean Prod 303:1–13

Eriksson A (2014) A discrete-event simulation approach to improve efficiency in stump fuel supply chains. Licensiate thesis, Swedish University of Agricultural Sciences, Department of Energy and Technology, Uppsala

Fan Y, de Kleuver C, de Leeuw S, Behdani B (2021) Trading off cost, emission, and quality in cold chain design: a simulation approach. Comput Indus Eng 158:107442. https://doi.org/10.1016/j.cie.2021.107442

FAO (2019). The state of food and agriculture 2019. Moving forward on food loss and waste reduction. https://www.fao.org/3/ca6030en/ca6030en.pdf. Accessed 10 Mar 2022

Figueira G, Almada-Lobo B (2014) Hybrid simulation–optimization methods: a taxonomy and discussion. Simul Model Pract Theory 46(11):118–134

Fikar C (2018) A decision support system to investigate food losses in e-grocery deliveries. Comput Ind Eng 117:282–290

Fikar C (2020) Model-driven decision support to facilitate efficient fresh food deliveries. Die Bodenkultur: J Land Manage Food Environ 71(1):1–9

Forster T (2013) Modellierung und Simulation von Getränkeabfüll- und Verpackungsanlagen unter Berücksichtigung von Energie- und Medienverbräuchen. In: Dangelmaier W, Laroque C, Klaas A (eds) Simulation in Produktion und Logistik 2013. HNI Verlagsschriftenreihe, Paderborn, Germany, pp 197–206

Fredriksson A, Liljestrand K (2015) Capturing food logistics: a literature review and research agenda. Int J Log Res Appl 18(1):16–34

Gäbel K, Tillman A-M (2005) Simulating operational alternatives for future cement production. J Clean Prod 13(13–14):1246–1257

Gharehyakheh A, Krejci CC, Cantu J, Rogers KJ (2020) A multi-objective model for sustainable perishable food distribution considering the impact of temperature on vehicle emissions and product shelf life. Sustainability 12(16):1–21

Goldratt E (1990) Theory of constraints. North River Press, Great Barrington

Gruler A, Fikar C, Juan AA, Hirsch P, Contreras-Bolton C (2017) Supporting multi-depot and stochastic waste collection management in clustered urban areas via simulation–optimization. J Simul 11(1):11–19

Haass R, Dittmer P, Veigt M, Lütjen M (2015) Reducing food losses and carbon emission by using autonomous control—a simulation study of the intelligent container. Int J Prod Econ 164:400–408

Haijema R, van Dijk N, van der Wal J, Sibinga CS (2009) Blood platelet production with breaks: optimization by SDP and simulation. Int J Prod Econ 121(2):464–473

Hatami-Marbini A, Sajadi SM, Malekpour H (2020) Optimal control and simulation for production planning of network failure-prone manufacturing systems with perishable goods. Comput Indus Eng 146:106614. https://doi.org/10.1016/j.cie.2020.106614

Holmgren J, Davidsson P, Persson JA, Ramstedt L (2012) TAPAS: a multi-agent-based model for simulation of transport chains. Simul Model Pract Theory 23:1–18

Hu G, Mu X, Xu M, Miller SA (2019) Potentials of GHG emission reductions from cold chain systems: case studies of China and the United States. J Clean Prod 239:1–11

ISO (2006a) ISO 14040:2006 environmental management—life cycle assessment—principles and framework. International Organization for Standardization, Geneva Switzerland

ISO (2006b) ISO 14044:2006 environmental management—life cycle assessment—requirements and guidelines. International Organization for Standardization. Geneva Switzerland
Ivanov D, Rozhkov M (2020) Coordination of production and ordering policies under capacity disruption and product write-off risk: an analytical study with real-data based simulations of a fast moving consumer goods company. Ann Oper Res 291(1):387–407
Jbira A, Jaoua A, Bouchery Y, Jemai Z (2018) Simulation based optimisation model for a joint inventory pricing problem for perishables. In Boukachour J, Abdelkader S, Alaoui AEH, Benadada Y (eds) Proceedings of the 4th international conference on logistics operations management. IEEE, Piscataway, NJ, pp 1–9
Jödicke G, Zenklusen O, Weidenhaupt A, Hungerbühler K (1999) Developing environmentally sound processes in the chemical industry: a case study on pharmaceutical intermediates. J Clean Prod 7:159–166
Johansson B, Johnsson J, Kinnander A (2003) Information structure to support discrete event simulation in manufacturing systems. In: Chick S, Sanchez PJ, Ferrin D, Morrice DJ (eds) Proceedings of the 2003 Winter Simulation Conference. IEEE, Piscataway, NJ, pp 1290–1295
Johansson B, Stahre J, Berlin J, Östergren K, Sundström B, Tillman AM (2008) Discrete event simulation with lifecycle assessment data at a juice manufacturing system. In: Proceedings of FOODSIM 2008 conference, Dublin, 26th–28th June, pp 165–169
Kather A, Voigt T (2005) Weihenstephan standards for the production data acquisition in bottling plants—Part 3: data evaluation and reporting. TU München, LVT, Freising
Kelton WD, Sadowski RP (2009) Simulation with ARENA, 5th edn. McGraw-Hill, New York
Ketzenberg M, Bloemhof J, Gaukler G (2015) Managing perishables with time and temperature history. Prod Oper Manag 24(1):54–70
Ketzenberg M, Gaukler G, Salin V (2018) Expiration dates and order quantities for perishables. Eur J Oper Res 266(2):569–584
Kheawhom S, Hirao M (2004) Decision support tools for environmentally benign process design under uncertainty. Comput Chem Eng 28:1715–1723
Kogler C, Rauch P (2019) A discrete-event simulation model to test multimodal strategies for a greener and more resilient wood supply. Can J for Res 49(10):1298–1310
Kohl J, Spreng S, Franke J (2014) Discrete event simulation of individual energy consumption for product-varieties. Procedia CIRP 17:517–522
Kouki C, Sahin E, Jemaï Z, Dallery Y (2013) Assessing the impact of perishability and the use of time temperature technologies on inventory management. Int J Prod Econ 143(1):72–85
La Scalia G, Micale R, Miglietta PP, Toma P (2019) Reducing waste and ecological impacts through a sustainable and efficient management of perishable food based on the Monte Carlo simulation. Ecol Ind 97:363–371
Law AM, McComas MG (1999) Simulation of manufacturing systems. In: Farrington PA, Nembhard HB, Sturrock DT, Evans GW (eds) Proceedings of the 1999 Winter Simulation Conference. IEEE, Piscataway, NJ, pp 56–59
Law AM, Kelton WD (1999) Simulation modeling and analysis, 3rd edn. McGraw-Hill, New York
Leithner M, Fikar C (2019) A simulation model to investigate impacts of facilitating quality data within organic fresh food supply chains. Ann Oper Res 314:529–550
Lin JT, Yeh KC, Sheu LC (1996) A context-based object-oriented application framework for discrete event simulation. Comput Ind Eng 30(4):579–597
Lin X, Negenborn RR, Duinkerken MB, Lodewijks G (2017) Quality-aware modeling and optimal scheduling for perishable good distribution networks: the case of banana logistics. In: Bektaş T, Coniglio S, Martinez-Sykora A, Voß S (eds) International conference on computational logistics. Springer, Cham, pp 483–497
Lind S, Krassi B, Johansson B, Viitaniemi J, Heilala J, Stahre J, Vatanen S, Fasth Å (2008) SIMTER: a production simulation tool for joint assessment of ergonomics, level of automation and environmental impacts. In de Vin LJ, Ng AHC, Thorvald P, Sullivan W G, Ahmad M (eds) Proceedings of the 18th international conference on flexible automation and intelligent manufacturing. University of Skövde, Skövde, Sweden, pp 1025–1032

Lütjen M, Dittmer P, Veigt M (2013) Quality driven distribution of intelligent containers in cold chain logistics networks. Prod Eng Res Devel 7(2):291–297
Ma AJ, Zhao HZ, Ren FZ (2010) Study on food life cycle carbon emissions assessment. Procedia Environ Sci 2:1983–1987
Mani M, Johansson B, Lyons KW, Sriram RD, Ameta G (2013) Simulation and analysis for sustainable product development. Int J Life Cycle Assess 18(5):1129–1136
McHaney R (1991) Computer simulation: a practical perspective. Academic press, San Diego
Melouk SH, Freeman NK, Miller D, Dunning M (2013) Simulation optimization-based decision support tool for steel manufacturing. Int J Prod Econ 141(1):269–276
Mustafee N, Powell J, Brailsford SC, Diallo S, Padilla J, Tolk A (2015) Hybrid simulation studies and hybrid simulation systems: Definitions, challenges, and benefits. In: Yilmaz L, Chan WKV, Moon I, Roeder TMK, Macal C, Rossetti MD (eds) Proceedings of the 2003 Winter Simulation Conference. IEEE, Piscataway, NJ, pp 1678–1692
Noordhoek M, Dullaert W, Lai DS, de Leeuw S (2018) A simulation–optimization approach for a service-constrained multi-echelon distribution network. Transp Res Part E: Logistics Transp Rev 114:292–311
Nunes MC, Nicometo M, Emond JP, Melis RB, Uysal I (2014) Improvement in fresh fruit and vegetable logistics quality: berry logistics field studies. Philos Trans Royal Soc A 372:20130307. https://doi.org/10.1098/rsta.2013.0307
Osterroth I, Klein S, Nophut C, Voigt T (2017) Operational state related modelling and simulation of the electrical power demand of beverage bottling plants. J Clean Prod 162:587–600
Persson D, Karlsson J (2007) Flow simulation of food industry production. Master thesis, Chalmers University of Technology, Department of Product and Production Development, Gothenburg
Pirard F, Iassinovski S, Riane F (2011) A simulation based approach for supply network control. Int J Prod Res 49(24):7205–7226
Polotski V, Gharbi A, Kenne JP (2021) Production control of unreliable manufacturing systems with perishable inventory. Int J Adv Manuf Technol 116(7):2473–2496
Psomas AN, Nychas GJ, Haroutounian SA, Skandamis PN (2011) Development and validation of a tertiary simulation model for predicting the growth of the food microorganisms under dynamic and static temperature conditions. Comput Electron Agric 76(1):119–129
Rijpkema WA, Rossi R, van der Vorst JGAJ (2014) Effective sourcing strategies for perishable product supply chains. Int J Phys Distrib Logist Manag 44(6):494–510
Ritchie H, Roser M (2020) Environmental impacts of food production. Our world in data. https://ourworldindata.org/environmental-impacts-of-food. Accessed 14 April 2022
Rohrer MW (2003) Maximizing simulation ROI with Auto-Mod. In: Chick S, Sanchez PJ, Ferrin D, Morrice DJ (eds) Proceedings of the 2003 Winter Simulation Conference. IEEE, Piscataway, NJ, pp 201–209
Safaei AS, Moattar Husseini SM, Z-Farahani R, Jolai F, Ghodsypour SH (2010) Integrated multi-site production-distribution planning in supply chain by hybrid modelling. Int J Prod Res 48(14):4043–4069
Santos R, Nunes J, Silva PDD, Gaspar PD, Andrade LP (2013) Computational tool for the analysis and simulation of cold room performance in perishable products industry. In: Proceedings of the 2nd IIR international conference on sustainability and the cold chain. Curran Associates, Red Hook, NY, pp 1–8
Sargent RG (2000) Verification, validation, and accreditation of simulation models. In: Joines JA, Barton RR, Kang K, Fishwick PA (eds) Proceedings of the 2000 Winter Simulation Conference. IEEE, Piscataway, NJ, pp 50–59
Scott EP, Heldman DR (1990) Simulation of temperature dependent quality deterioration in frozen foods. J Food Eng 11(1):43–65
Sel Ç, Bilgen B (2014) Hybrid simulation and MIP based heuristic algorithm for the production and distribution planning in the soft drink industry. J Manuf Syst 33(3):385–399

Shamsi AA, Raisi AA, Aftab M (2014) Pollution-inventory routing problem with perishable goods. In: Golinska P (ed) Logistics operations, supply chain management and sustainability. Springer, Cham, pp 585–596

Solding P, Thollander P (2006) Increased energy efficiency in a Swedish iron foundry through use of discrete event simulation. In: Perrone LF, Wieland FP, Liu J, Lawson BG, Nicol DM, Fujimoto RM (eds) Proceedings of the 2006 Winter Simulation Conference. IEEE, Piscataway, NJ, pp 1971–1976

Soto-Silva WE, Nadal-Roig E, González-Araya MC, Pla-Aragones LM (2016) Operational research models applied to the fresh fruit supply chain. Eur J Oper Res 251(2):345–355

Teller C, Holweg C, Reiner G, Kotzab H (2018) Retail store operations and food waste. J Clean Prod 185:981–997

United Nations Environment Programme (2021) Food waste index report 2021. https://wedocs.unep.org/bitstream/handle/20.500.11822/35280/FoodWaste.pdf. Accessed 1 March 2022

Utomo DS, Onggo BS, Eldridge S (2018) Applications of agent-based modelling and simulation in the agri-food supply chains. Eur J Oper Res 269(3):794–805

Van Der Vorst JG, Tromp SO, Zee DJVD (2009) Simulation modelling for food supply chain redesign; integrated decision making on product quality, sustainability and logistics. Int J Prod Res 47(23):6611–6631

Vicino DA (2015) Improved time representation in discrete-event simulation. Ph.D. thesis, Université Nice Sophia Antipolis, Informatique, Sophia Antipolis

Voigt T (2004) Neue Methoden für den Einsatz der Informationstechnologie bei Getränkeabfüllanlagen. Ph.D.thesis, TU München, Wissenschaftszentrum Weihenstephan, Weihenstephan

Wohlgemuth V, Page B, Mäusbacher M, Staudt-Fischbach P (2004) Component-based integration of discrete event simulation and material flow analysis for industrial environmental protection: a case study in wafer production. EnviroInfo 1:303–312

Wu W, Beretta C, Cronje P, Hellweg S, Defraeye T (2019) Environmental trade-offs in fresh-fruit cold chains by combining virtual cold chains with life cycle assessment. Appl Energy 254:113586. https://doi.org/10.1016/j.apenergy.2019.113586

Zhang Y, Lu H, Zhou Z, Yang Z, Xu S (2021) Analysis and optimisation of perishable inventory with stocks-sensitive stochastic demand and two-stage pricing: a discrete-event simulation study. J Simul 15(4):326–337

Chapter 7
Renewables

Cedric Schultz, Martin Rösch, and Lukas Bank

Abstract In order to reduce greenhouse gas emissions and energy costs, renewable energy sources are of growing importance for manufacturing systems. This chapter gives an outline of different renewables, the complexities that arise from their use, and explains how simulation studies can be applied to address the interactions between manufacturing systems and renewables. To give examples of the challenges of integrating renewables in manufacturing, two applications are presented for simulation in the design and evaluation of manufacturing control strategies under constraints of renewable energy sources. The simulation models include several forms of renewable energy sources, manufacturing resources as well as an energy storage. In these applications, the simulation serves two different purposes. In the first study, the simulation is used to generate training data for an AI algorithm based on multi-agent reinforcement learning. This algorithm is applied to a closely linked manufacturing and energy system in order to balance energy supply and demand. The second application presents how a customized energy tool for a commercial material flow simulation can be used to generate high-resolution load forecasts applying scenarios for order scheduling. Based on the simulation result, the best scheduling alternative for a given energy supply by renewables can be chosen.

C. Schultz (✉)
TARGUS Management Consulting AG, Dechenstraße 7, 40878 Ratingen, Germany
e-mail: cedric.schultz@targusmc.com

M. Rösch
Bayernwerk Netz GmbH, Lilienthalstraße 7, 93049 Regensburg, Germany
e-mail: martin.roesch2@bayernwerk.com

L. Bank
Drakestraße 45, 12205 Berlin, Germany
e-mail: lukas.bank@web.de

S. Wenzel et al. (eds.), *Energy-Related Material Flow Simulation in Production and Logistics*, https://doi.org/10.1007/978-3-031-34218-9_7

7.1 Introduction

News and events of the past decades have increased customer awareness for decarbonization and, thereby, increased the attention of the society as well as of companies on CO_2 reduction or even neutrality (United Nations 2021). The transformation of the energy systems towards renewable energy sources is seen as a key element to reduce greenhouse gas (GHG) emissions and inhibit anthropogenic climate change (European Commission 2021). In the year 2020, more than 19% of electricity consumed in Germany was generated by renewable energy sources. This amounts to a total of 455 billion kWh electricity and represents an average share compared to other EU countries (Fig. 7.1) (Umweltbundesamt 2021; Eurostat 2021). In 2017, the share amounted to 9.9% in the U.S., to 12.8% in China, and to 17.3% worldwide (International Energy Agency 2021; International Energy Agency et al. 2020).

Globally, costs for renewable energy sources have declined significantly in the past decade. For example, the costs for utility-scale solar power has decreased by 85% from 2010 to 2020 and costs for wind power by roughly 50% (IRENA 2021). At this point, renewables can be economically competitive with fossil fuels and, as shown in Table 7.1, may facilitate cost savings as well as a reduction of GHG emissions (IRENA 2021). Nevertheless, this potential is still highly dependent on the individual region in question (International Energy Agency 2020) and has to be analyzed in detail for individual applications.

Due to the multitude of stochastic factors influencing the economic viability of renewables in a specific case—especially regarding the variability of energy demand and weather influences in a region—simulation models can be helpful and often necessary to determine the feasibility and potential outcome of an application of renewables. Such simulative evaluations can be carried out at a level that covers a whole economy as well as on a plant or product level (see Sect. 7.3).

There is a multitude of technologies available for renewable power generation. But in general, discussions are focused on either hydroelectric power, wind power, solar power, or biomass. Since international financing for renewables in developing

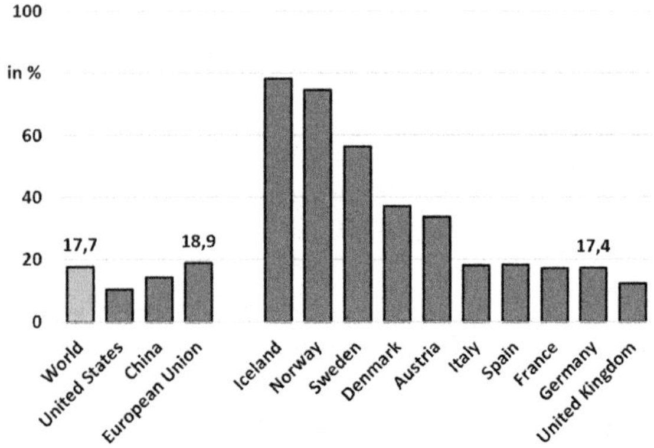

Fig. 7.1 Share of renewable energy in final energy consumption (Umweltbundesamt 2021; Eurostat 2021; International Energy Agency 2021; International Energy Agency et al. 2020)

Table 7.1 Potential annual savings and CO_2 emissions reductions by replacement of coal with solar and wind power (IRENA 2021)

	Coal capacity with higher operating costs than new solar and wind		Annual saving from replacing coal with new solar and wind	Annual CO_2 emission reductions
	(GW)	Including integration costs (GW)	(USD billion/year)	(Mto CO_2/year)
Germany	28	28	3.3	99
United States	188	149	5.6	332
World	1137	810	32	2973

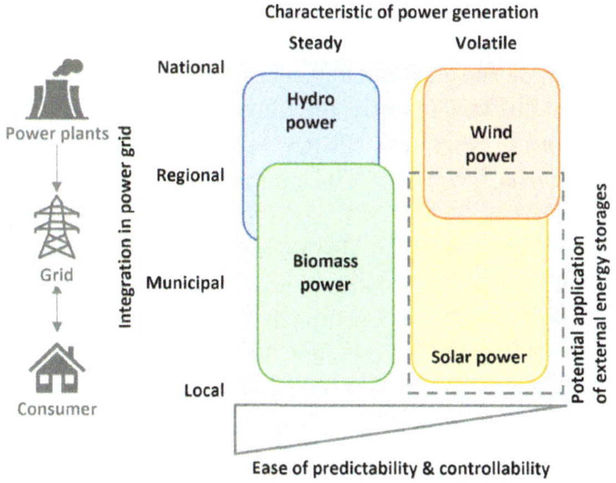

Fig. 7.2 Characterization of renewables

countries has been increasing continuously, a further increase of the share of renewables—especially solar power—can be expected worldwide (United Nations 2019). As shown in Fig. 7.2, these energy sources differ in the way they supply energy and in their application in the energy grid.

7.2 Scope and Objectives

Hydroelectric power plants are typically large in scale and, therefore, mainly used on a national or regional level of the energy grid. The waterflow providing the energy can be readily controlled and, thus, predicted. Wind power is generated by wind turbines. These turbines are also most economical in larger scales. While local applications exist, the focus of wind power is seen on a regional level. Due to the stochastic nature of wind, power generation of wind turbines fluctuates significantly over time

(Kaltschmitt et al. 2020; Konstantin 2013). Solar power can be collected by photovoltaics on a wide range of scale including regional solar farms as well as photovoltaic (PV) installations for individual buildings on a local level. Power generation by PV shows a characteristic curvature over the course of a day, but predictability of the actual amount of power generated is limited (Kneiske and Hoffmann 2014). Biomass power plants essentially burn regrowing and ideally sustainably managed biomass, such as wood pallets, farm waste, etc., to power steam generators and turbines. Technically, the process is scalable and can be used on a local level. However, larger plants tend to be more efficient. The turbines are easily controllable and the amount of power, therefore, predictable (Konstantin 2013).

Manufacturing companies may be motivated to consider renewables for two main reasons: carbon neutrality and energy costs. GHG emissions can be avoided when renewable energy sources are used and, therefore, the carbon footprint of produced goods can be decreased. This reduction can be achieved either on a balance sheet by buying an increased share of energy produced by renewables on a national level or by directly producing energy locally through renewables (Kneiske and Hoffmann 2014). Currently, the European Commission is working on new Sustainability Reporting Standards (ESRS), which define how corporations have to account their carbon footprint and GHG emissions (European Financial Reporting Advisory Group 2022).

Energy costs may also be influenced by renewables on a national and on a local level. Given the volatile characteristics of some renewables, the growing share of these energy sources results in increasing effort to stabilize energy grids and balance energy supply and demand on a national scale. Consequently, time-of-use tariffs and peak-load management may be even more relevant in future and incentivize companies to focus their energy demand on off-peak hours. In the future, energy storages could also potentially be used for grid stabilization (International Energy Agency 2020). Additionally, companies may use renewables themselves locally to cover a part of their electricity demand. As some technologies (e.g., photovoltaics) have reached grid parity, i.e., they can produce electricity at costs lower than that of grid electricity, companies may directly reduce energy costs this way.

Whether renewables are applied to reduce carbon footprint or energy costs (Table 7.2), the result is that electric energy is no longer viewed as an after-the-fact cost factor, but as time-dependent and limited resource whose use must be planned carefully. It might, therefore, be necessary to assess, how much energy from which energy sources is used during stages of manufacturing over the course of time. Production might even be changed due to time- or load-based incentives. Additionally, employing conversion factors to translate usage of electricity into emissions of GHG, simulation can evaluate the overall carbon footprint of produced goods. Because of the interrelations between manufacturing and energy systems and the partially time-dependent factors of energy supply, simulation is a well-suited approach for this kind of assessments.

Table 7.2 Scope of application for renewable energy

Scope	Nationally	Locally
Carbon footprint	Balance sheet (out of scope)	Onsite generation through renewables
Energy costs	Incentives (time-of-use, load management)	Onsite generation through renewables

7.3 KPIs and Assertions from Simulation

Interactions between renewable energies and a production or logistic system are often complex because of the interdependencies among material flow, energy supply, and the energy demand by individual electric consumers. Therefore, simulation models have been proposed and used for many different applications including—but not limited—to the following categories:

Microeconomic sustainability metrics are applied for the sustainability evaluation of entire energy systems, economies, or industries. Studies within this category often involve large-scale problems of how the adoption of certain energy policies or implementation of renewable energies might affect energy supply and costs, economic output, and GHG emissions of entire regions or countries, often using system dynamics approaches. Examples can be found in Mazhari et al. (2011), Jain et al. (2012), Robalino-Lopez et al. (2014), Shih and Tseng (2014), Zahraee et al. (2016), and Kelly et al. (2019).

Operational sustainability metrics are used to evaluate the sustainability of certain manufacturing systems, products, or technologies. On a smaller scale, simulation studies are carried out to assess the impact of renewable energies on specific production environment. Typical problems examine how an increased share of renewables influences the overall GHG emission, manufacturing costs and performance throughput at an individual manufacturing site or for an individual product. Examples can be found in Roedger et al. (2021) and Materi et al. (2021).

The *sustainability of energy supply chains* is measured to evaluate renewables or the renewable energy supply chain. Within this application, simulation models are used to determine, how sustainable and economically viable renewable energy sources are over the long term. This can include all steps of energy conversion and transmission between the source and the consumer (Energy Supply Chain). In this context, the total reduction or prevention of GHG emissions as well as the overall life cycle cost (LCC) are important KPIs. Approaches might utilize Monte Carlo Simulation (Jeon and Shih 2014) or System Dynamics (Saavedra et al. 2018).

The *design of energy supply systems* is in the focus of planning energy systems and distributed energy supply. A significant share of simulation studies are focused on the optimal planning and dimensioning of the energy supply for a region or an individual manufacturing site. Problems often include renewables with their volatile power generation characteristic, multiple consumers with an assumed energy demand, as well as conventional energy sources. Typically, KPIs such as the overall system costs, the levelized cost of energy (LCOE), the utilization of renewable sources, the loss of load probability (LLP) and the overall GHG emissions are studied. Examples are found in Hollmann (2006), Sanders et al. (2012,) Taboada et al. (2012), Reddi et al. (2013), Binbin et al. (2017), Thornton et al. (2018), Woltmann et al. (2018), Schulze et al. (2019), Tongdan et al. (2020), and Oladeji et al. (2021).

Energy-conscious manufacturing operations are targeted for the design and validation of energy and manufacturing planning as well as related control strategies and algorithms. Typically, for these applications a production system including multiple machines as electric consumers is modeled in combination with the energy supply by renewable sources. Targets are usually the optimization of manufacturing costs including costs for electricity or the alignment of production orders and, therefore, energy demand with constrained energy supply while maintaining throughput and utilization of the manufacturing system. In some cases, additional consumers (e.g., building installations, ventilation, etc.) are also included. Examples are found in Schultz et al. (2015), Weinert and Mose (2016), Beier et al. (2016), and Beier (2017).

In the following sections, simulative approaches to design and validate control strategies that align manufacturing with renewable energy supply will be further explored. Additional simulative applications can be found in Moon (2016).

7.4 Data Requirements

Data requirements for material flow simulation of renewables are highly dependent on the individual scope and application of the model. In many cases, the focus of the model is to find an optimum between production needs, i.e., preferably continuous and unlimited energy availability, and a discontinuous or even intermittent energy supply by renewables. As such, general requirements for the description of the production system apply (see Chap. 2).

In addition to the production system, energy supply and demand need to be modeled. Depending on the scope of the application, the model may include different electrical consumers such as manufacturing resources, building installations, or even electric vehicles within the production system. Their energy demand is often represented by a state-based modeling approach (Thiede 2012). Thereby, each relevant state of a consumer (e.g., "Off", "Idle", or "Producing") is characterized by a specific energy demand. This energy demand is most often modeled as a constant average value in kW, but may also be modeled as more complex load profiles per state (Weinert et al. 2011).

Depending on the scope of the application (see Table 7.2), the energy supply side includes renewables directly or indirectly through incentive-based pricing. When local power generation by renewables is studied, these energy sources are typically modeled as time series representing their power feed over the course of a shift or a day. These time series are either based on historic data collected on site or on reference data for similar installations of renewables (Beier 2017). In any case, volatility needs to be taken into account (Fig. 7.2).

Renewables that are integrated into the electric grid on a national level only have an indirect effect on individual consumers. Due to their volatility, there might be times with a high supply of cheap electricity (actually straining the grid) and times where supply is unexpectedly low. These circumstances are typically addressed and modeled by time-based pricing of electricity. Therefore, the time series required for the simulation specify the costs of electricity over the course of the day instead of actual power generation. In addition—and independent of whether renewables are considered or not—there are usually thresholds for maximum peak loads in the contracts with energy providers. Surpassing these peak loads leads to sharp increases in yearly energy costs.

If the impact of renewables on the CO_2 footprint of the company, production or product is studied, this can also be represented by a CO_2 coefficient per kWh of the local energy supply as well as of the energy supplied by the grid.

7.5 Challenges for the Simulation Models

Even though specific challenges in simulating renewables depend on the individual application, common problems to be addressed in the model often include the volatility and time resolution of renewable energy sources. As described in Sect. 7.1, wind and solar power show volatile behavior making prognoses based on historical data challenging (Tao et al. 2010; Xu et al. 2012; Gonzalez Ordiano et al. 2017; Oneto et al. 2018).

If renewables are modeled on a larger timescale of weeks, months, or longer, for example if the CO_2 footprint of production is to be studied, this volatility can be addressed by averaging. However, if the application is based on a shorter timescale of hours or minutes, significant uncertainty and forecast error have to be considered.

The same is true for the demand side of the model, which has to match the supply side. Applications in the field of production planning and control typically ask for high resolution data, i.e., minutes, of energy demand depending on operating states of consumers, products, or manufacturing orders. Load peaks are especially demanding, since they usually result from a superposition of different energy consumers within seconds or minutes, and continuous energy profiles might be required for proper modeling. For many practical applications these data are not readily available and obtaining demand data for all consumers might proof challenging, since large-scale energy measurement has not become industry standard today. Even if data for energy demand are collected in manufacturing, they are often handled in systems and databases separate from the manufacturing operation itself and need to be linked to manufacturing jobs.

When integrating energy into a material flow simulation, the challenges are the different requirements for the simulation. A joint consideration can only be achieved with substantial effort (Schlegel et al. 2013). Material flows in discrete manufacturing can be well represented by discrete event simulation, but this technology is only suitable to a limited extent for representing continuous energy flows. To counter this

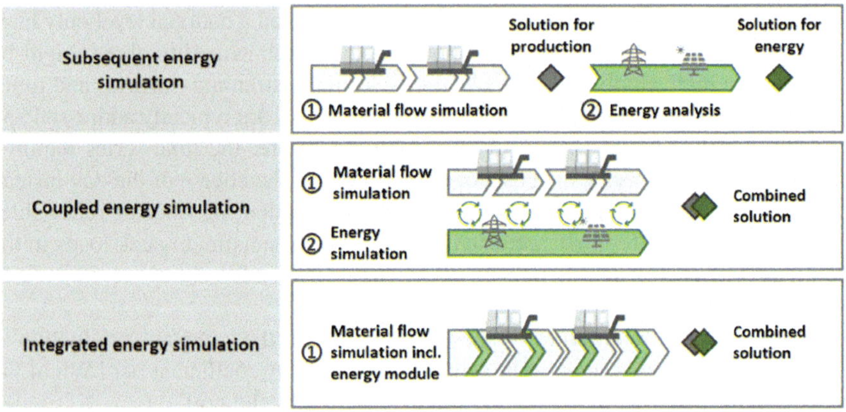

Fig. 7.3 Paradigms for a joint simulation of material and energy flow

problem, the three essential paradigms shown in Fig. 7.3 have been developed (Thiede 2012). These paradigms were already introduced as part of the morphological box in Sect. 1.2.2 regarding the timing and architectural approach of the energy evaluation.

Subsequent energy simulation This approach is carrying out the material flow simulation without considering energy. Subsequently, the energy is analyzed in separate simulation model or other tool based on the result of the material flow simulation.

Coupled energy simulation This paradigm is coupling a material flow simulation using a discrete event simulation with a continuous simulation tool for the representation of the energy flows in order to achieve a simultaneous consideration of the interactions and to use the appropriate tool for the representation in each case. An example can be found in Junge (2007).

Integrated energy simulation The energy flow is integrated in the material flow simulation, e.g., by discretizing the energy flow, to achieve an integrated view between material and energy flows.

The first paradigm can be implemented with relatively little effort and is suitable if the effects of the material flow on the energy flows are to be considered and no adjustments of the material flow based on energy variables are to take place. The second paradigm uses corresponding specialized applications for both energy and material flow. However, the coupling is complex, as two models have to be created and the corresponding know-how has to be available. The third paradigm uses material flow simulation and integrates an energy flow consideration. The energy flow must be discretized for a discrete event simulation. Therefore, accuracy is lost in the energy consideration. The choice of the paradigm, thus, depends on the question that is to be answered by the simulation. A consideration of the individual advantages and disadvantages has to be considered individually.

7.6 Applications

For the better understanding of the preconditions and potential outcomes of simulation and to illustrate the discussion of the chances and challenges of simulation technology, this section introduces two use cases from different perspectives. The first case (Sect. 7.6.1) studies the cost saving by avoiding peak loads in the public power supply of a manufacturing system by exploiting batteries and renewable energy sources. The driving idea is that with a smart control strategy, the production could be levelled in its power consumption, and peaks of the volatile renewable energy efficiently used. The second use case (Sect. 7.6.2) is a medium-sized manufacturer of bevel gears and related parts, which during their production undergo hot forming and thermal post-treatment. The high power demand is partially satisfied by a PV system, and suitable strategies are sought to exploit the volatile energy input from the PV as completely as possible in the production, and also to schedule the consumption of electricity from the grid at time windows where the fluctuating prices are low. Simulation helps to identify the right measures, especially when disturbances force to change the original scheduling.

7.6.1 Integration of Self-sufficient Energy Supply in Manufacturing

This section describes the implementation and usage of a simulation model from a medium-sized metal working company. The roughly 120 employees at this production site focus mainly on the precision finishing of metal parts for the automotive sector as well as tool manufacturing. Hence, the manufacturing system mainly consists of several production resources on the energy demand side and focuses on the integration of power supply based on renewable energy sources and a battery system. The resulting simulation model is applied to gather data and scenarios to train a control system based on artificial intelligence, which is finally able to control the complex system in a holistic and efficient way.

7.6.1.1 Initial Situation and Goals of the Study

Controlling a manufacturing system is a complex control task. Regarding the integration of renewable energy sources, their power generation is volatile and only partly controllable. Thus, for an optimal use of this energy, the electricity demand of the manufacturing system also needs to be aligned with the availability. This results in an even more complex control problem, which entails a need of new and more powerful algorithms, especially as stochastic events require a highly reactive system. Recent breakthroughs in the field of artificial intelligence (AI) open new possibilities to tackle such challenging control tasks. However, for training such an AI, in general a huge amount of training data is required, which often cannot be acquired in a real-world system. Instead, a simulation can be used. Eventually, the overall goal is to minimize both, energy and manufacturing costs in the same way.

In this context, the mentioned AI is applied for an energy-oriented production control, implemented for a medium-sized company (Roesch et al. 2020). Its production site consists of a complex energy system with the following elements (Fig. 7.4):

- Battery storage (BS)
- Combined heat and power plant (CHP)
- Photovoltaics system (PV)
- Public grid connection (PGC)

On the electricity consumer side, there are mainly five production resources for metal working, who cause together roughly 60% of the total electricity consumption. Defined manufacturing jobs of several types with specific durations are assigned to these resources, resulting in a discrete manufacturing system. Between two jobs of different types, a machine setup is required. The other 40% of electricity are consumed by various smaller devices and machines as well as elements of building services.

On the energy side, it is highly attractive to consume the self-generated electricity from PV and CHP, as the specific costs especially from PV are lower than electricity supplied from the public grid. In addition, high peak loads from the public grid should be avoided. The reason is, that in the German electricity grid, the average load in 15-min time intervals is decisive for the generally expensive charges for peak power consumption. On the production side, jobs should be finished in time. The number of job types to be machined is predefined by the previous capacity planning process, which is conducted by a custom planning tool using a heuristic optimization algorithm based on simulated annealing (van Laarhoven and Aarts 1992) and should be out of scope for this book. The electricity and production system is brought together, as the manufacturing resources are consuming the energy and, thus, have the potential to contribute to minimizing energy costs by an intelligent schedule.

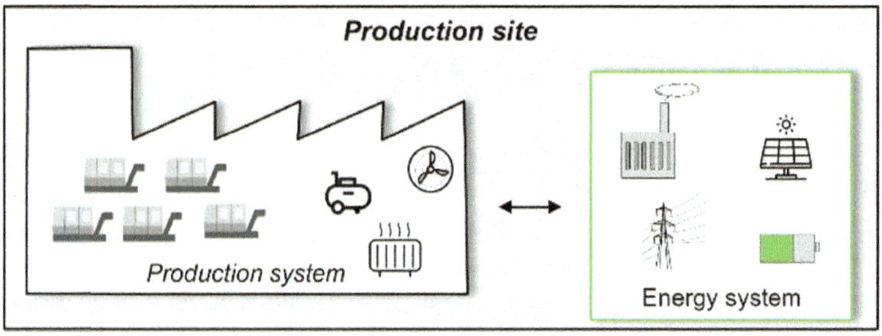

Fig. 7.4 Schematic system overview for the metal working company

Under these circumstances, the objective is to create a simulation model that displays the logic and interdependencies of both, the production and energy system to generate training data for the AI. Based on this, the latter shall be applied to provide a suitable and holistic short-term control strategy to minimize electricity and manufacturing costs in the same procedure. Therefore, it is important for the simulation to evaluate the quality of control actions taken by the AI system, creating an objective function that allows for assessing the quality of a control strategy.

7.6.1.2 Procedure of the Simulation Project

The process of creating the simulation model was inspired by the procedure described in the VDI guideline for simulation of systems in logistics, material handling and production (Verein Deutscher Ingenieure 2014). In the following, the main specifications and tools are summarized.

In order to understand the dependencies and cohesion of the involved system elements, a detailed analysis of the real-world system was made. Besides the production data—like the required energy, time, and resources for discrete manufacturing tasks—also the overall energy consumption data were gathered. As some detailed energy data were not recorded by the installed energy management system, additional electricity measurements were arranged temporarily. During the following modeling step, the system model has been created based on a modular concept and the data were preprocessed and prepared. This was performed by a newly programmed time-step-based simulation environment using Python and C, inducing a bigger effort to create all models from scratch compared to using an existing simulation program. However, as computational performance is essential for using a model for training an AI, this approach has been selected. Because the open source library RLlib (RAY 2022) was applied for the AI part, the model could be easily integrated.

After iterative verification, beginning on single element level and advancing to the entire system, the simulation results were evaluated as well as parameters adjusted and defined. Thereby, the existing real-world data regarding energy consumption and production plan of the entire system were used. Finally, the simulation has been integrated in the AI training pipeline. Thereby, the simulation can be executed providing an environment where the AI control can interact and learn on the experience it is gathering during its training. To assess the quality of the taken control actions, an additional evaluation function is created that depicts the resulting energy costs and quantifies the attainment of general logistic objectives of the production system. As an AI algorithm, a multi-agent reinforcement learning approach was applied (Gabel 2009). The resulting architecture is outlined in Fig. 7.5.

After presenting the procedure on a very high level, this section provides a deeper understanding of the major characteristics of the simulation model, especially regarding the linkage of energy and production system.

Discretization In order to limit the computational effort, the given model is discretized over time. Although the energy flow is a continuous process, the energy

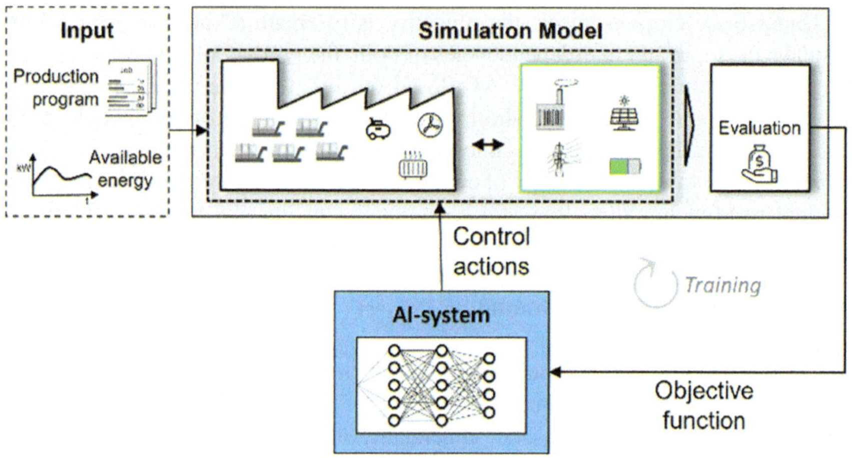

Fig. 7.5 System architecture

consumption is not modeled by a continuous function, but a discrete time simulation is applied. This concludes in the approach to choose time steps as long as possible and as short as necessary. Regarding the overall goals of the given study, a major time constraint is the avoidance of peak loads, which are accumulated in intervals with a length of 15 min (see Sect. 7.6.1.1). Therefore, the AI-based controller needs to react within a 15-min interval to prevent conceivable peaks. Besides, the length of a single time step should be a common denominator of the relevant billing interval of 15 min. From the production side, the jobs assigned to the manufacturing resources take between 20 and 40 min. Consequently, three minutes were defined as the length of a time step, resulting in each 15-min interval consisting of five consecutive time steps. Doing so, the machining time of jobs had also to be rounded to a multiplicative of three, resulting in a duration, e.g., of 21 instead of 20 min. This leads to some blur of the simulation model. This is, however, acceptable in favor of computational expense and performance of the system.

Focus and system boundaries To further reduce the system complexity and focus on the most relevant aspects, some important measurements are taken, which are briefly explained in the following.

With respect to the production system, the simulation focus was put on the resources with the highest energy consumption. Therefore, only the mentioned five manufacturing resources are modeled in detail, which in total cause roughly 60% of the total energy consumption. As the rest of the energy demand is caused by several smaller consumers with poor data quality and less direct dependence on the production schedule, the analysis concluded in not displaying these parts in detail (Fig. 7.6). The consumption of these consumers was rather aggregated and modeled as not controllable, regarding it as an unswayable input. On a technical level, this input consists of a time series based on historical data of the real system.

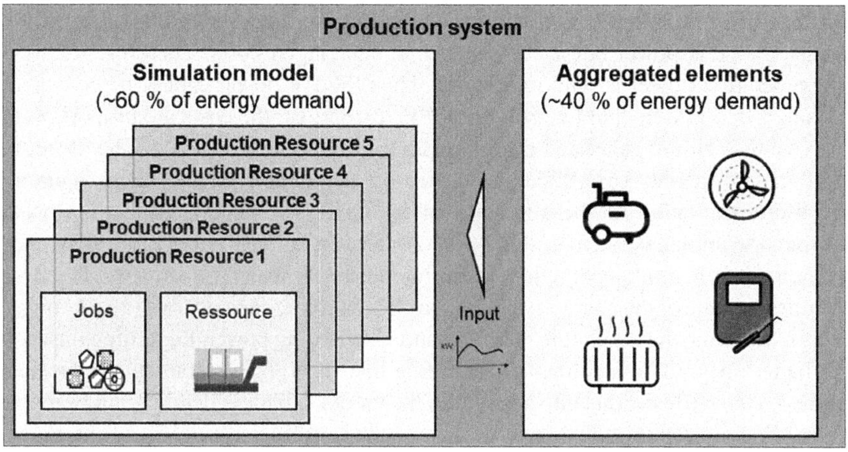

Fig. 7.6 Overview modeling concept of the production system

Another simplification was made for the energy consumption of jobs machined in the production resources. In the given production system, there is a limited number of job types, with each job of a type having the same energy consumption and machining time. In fact, the energy demand is fluctuating during the machining process. However, cumulated to a longer time interval of several minutes, the consumption is roughly constant during the whole machining process. Thus, for simplification the jobs are modeled with a constant power consumption. However, this leads to a decent blur regarding the short-term fluctuation of energy demand. In analogy to the technical side, the energy demand and availability are cumulated within a time interval of three minutes anyway, as this seems acceptable. Due to this averaging within a time interval, the impact and risk of, e.g., very short but high demand peaks can be neglected.

The simulation system is furthermore based on various system elements, each representing an energy consumer or generator. Thus, every manufacturing resource represents a consumer, whereas CHP, PV, and PGC are the three generators. The BS plays a special role, as this system element can act as both consumer and generator.

Connecting material and energy flow One major challenge of the simulation study results from the two flows of energy and material, which present different physical dimensions, but at the same time are strongly connected. On the part of the energy system elements (e.g., PV, CHP, and BS), there is no direct linkage to the material flow and manufacturing objective. On the part of the production resources, however, a job entails a specific machining time and energy consumption in the same way. Consequently, both aspects have to be modeled. For implementation, both parts were programmed in a modular way using Python, which leads to the benefits mentioned above regarding the computational performance.

Having a closer look at the modelling of the energy system, the PV and CHP are modeled quite differently. The maximum power in every time step of a PV is defined by the current weather conditions. Based on this, the electricity output can only be adjusted to less power but at the same time unchanged costs. Therefore, reducing the generation power is only reasonable when the PV power cannot be consumed, stored, or fed into the public grid. For CHP, in contrast, the costs are proportionally increasing with the generated power, as fuel needs to be consumed. Therefore, the CHP can be flexibly controlled and turned off completely, in case there is enough electricity available. As the efficiency of the CHP is in general depending on the current operational point, it is often not economically efficient to operate a CHP under a defined power limit, leading to another boundary of the control strategy. The BS as the third module of the energy system has a characteristic capacity rate (also known as C-rate) to define the maximum charging and discharging power. Regarding the costs, the battery degradation defines the resulting live span and economical depreciation of the system. To take this into account in the model, a battery degradation is applied online as part of the simulation, to determine the resulting costs of every charging or discharging action.

To model and assess the electricity costs, all generated and consumed electricity of each system element is cumulated in an electricity pool with an internal balance sheet for consumed and generated electricity, which is indicated in Fig. 7.7. As the billing interval for the considered electricity pool is 15 min, the pool was filled every time step (3 min). However, the balance sheet was interpreted and then cleared after every 15 min. In addition, the generated energy is assigned to specific costs, based on its origin. For example, the electricity generated by CHP has other costs than the consumed power from the public grid. In order to display the impact of the material flow on the production, a synthetic cost overall function $OC_n(t)$ is created (Eq. 7.1). On the one hand, it consists of costs for the setup time CS_n for every job n and additional labor force LC_n in case of extra hours, and on the other hand of time-dependent costs $TC_n(t)$ entailed by missed due dates. Eventually, the material flow as well as the electricity flow can be monetarily assessed, providing an objective function for the complex system.

$$OC_n(t) = \sum TC_n(t) + \sum CS_n + LC_n \tag{7.1}$$

Data generation and data augmentation The overall purpose of the simulation study is to provide a suitable environment for an AI algorithm to generate and aggregate training data. For the training, data of several thousand working days are required. This prerequisite forms a major challenge of the study, as there are no real-world data in this extend. Consequently, the given limited data from the real-world production site had to be augmented, based on randomization factors for PV generation, stochastic events in the production leading to breakdowns, energy demand from other consumers than the modeled resources, and the composition and amount of jobs to be manufactured. The determination of these randomization factors is crucial for the success and ability for generalization of the AI algorithm. As the generated data should be realistic, but also provide different scenarios that did not yet occur but are possible in the future, stochastic models are applied. The necessary complexity of

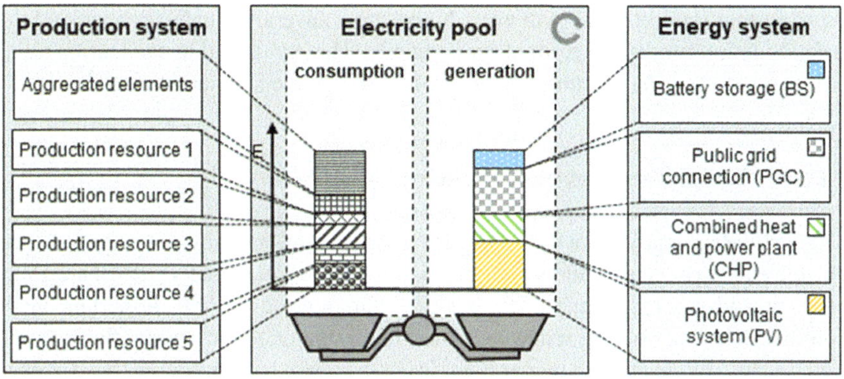

Fig. 7.7 Electricity pool concept

these models depends on the magnitude and number of influencing factors, which have an impact on the data series to be augmented. As an example, for the generation of PV data a simple Gaussian distribution is applied. To generate consumption data for the energy demand of other consumers than the modeled production resources, a complex neural network is trained based on several influencing factors like the set of jobs to be produced, the time of day, and the season.

7.6.1.3 Results of Experiments and Evaluation

The synthetic data generated by the simulation are used to train the AI algorithms. Thereby, the AI controls the strategy of every system element, defining, e.g., whether to charge or discharge the battery and which job to process at every production resource at the same time. The training process is computationally expensive. In this context, the simplifications and chosen discretization prove to be an important contribution for a computational-efficient simulation. This results in being able to reduce the training duration to several hours, which requires roughly 3,000 simulated training episodes (each episode representing one working day). After the successful training on the synthetic data, the AI is applied on several episodes of the real-world production system data for validation. Doing so, the AI algorithm proved to have learned a generalized strategy that provides robust and good results for controlling the overall system.

After training, the AI is able to take decisions within seconds, which provides a highly reactive control strategy enabling fast reactions on unforeseen events. Thus, the whole system including manufacturing resources and the energy modules can be controlled in a holistic and highly efficient way. Compared to a conventional rule-based heuristic, which provides a comparable reactivity, the trained AI algorithm could halve the resulting overall production costs. At the same time, the application of a metaheuristic optimization approach concludes that there is still potential for the

AI to further reduce the overall costs by on third. Overall, it can be concluded that the simulation fulfills its required functionality. The AI algorithm can learn a robust strategy based on the training scenarios provided by the simulation model.

7.6.1.4 Benefits and Lessons Learnt

The simulation study shows that besides a simple and modular concept, the focus on relevant aspects as well as suitable simplifications essentially contribute to the efficient and successful approach. These findings are especially important in the context including volatile renewable energies of material and energy flow simulation, as these entail an even higher complexity compared to simulation only focusing on controllable energy supply units. In general, regarding the great expansion of applications using AI, simulation is a powerful technology to generate sufficient and realistic training data, providing an important contribution for future breakthroughs and optimized control strategies for integrating renewable energy sources in an efficient way. On a system level, the presented control strategy can further be extended on more complex setups on production side (e.g., more manufacturing resources and additional constraints) as well as integrating trading on the short-term electricity markets on the electricity side.

7.6.2 Load Forecasting of the Self-Production Rate of Solar Power

The company where this simulation study was carried out manufactures, among other things, bevel gears. These are installed in the chassis of vehicles. The company is a medium-sized enterprise that can be described as a Tier-2 supplier and has several plants. In the simulation study, one production area of a plant was considered, which is responsible for the main power demand of the site. The production area is located in a hall where bevel gears are produced by hot forming. In order to obtain the desired material properties, a heating process follows the hot forming. Energy is required at the hot forming machines for heating the material and for mechanical forming. The forming process is also followed by thermal post-treatment, which consists of several furnaces. The site has a photovoltaic (PV) system. The aim is to use as much of the electricity generated there as possible and to purchase additional electricity cheaply on the basis of a plan. In the event of deviations from the production plan, suitable strategies are to be evaluated in the simulation.

7.6.2.1 Initial Situation and Goals of the Study

Energy costs have a significant share in production-related costs. Today, production and energy management tasks are usually considered separately in industrial applications, losing the knowledge about the connections between the production program and the energy demand. In processes with energy-intensive equipment, this can lead to problems, if, e.g., a load peak occurs due to parallel operation of this equipment. Load peaks have a negative impact on grid stability and are, therefore, subject to penalty payments. In order to avoid these payments, industrial companies sometimes resort to drastic measures and switch off plants for a short time, with corresponding negative effects on production. A load forecast is relevant for the integration of self-generation plants into factory operations to adjust consumption depending on the expected generation. The goal is to consume the company's own electricity as completely as possible. Accordingly, it can be useful to schedule consumption-intensive production processes during periods of high electricity availability. In this respect, on the one hand, appropriate planning is important. On the other hand, in the event of changes, appropriate measures must be developed to respond to these changes. Changes to the plan can arise from electricity demand, e.g., due to machine failure or from energy production, if production is lacking behind the expectations. This is especially the case when electricity generation is weather-dependent renewable energy. Forecasts are of high quality only with short time horizons, but they do not protect against unforeseeable fluctuations in actual generation. Then, the fluctuations in self-generation must be adjusted either by adjusting the company's own demand or by drawing from the grid. The adjustment of the energy demand can lead to a reduction of the output depending on the amount of load reduction and a corresponding decision should be secured. The balancing via the power grid can additionally create load peaks.

The use case takes up this problem and shows how a load forecast is carried out with the help of simulation to recognize imminent load peaks and to develop suitable strategies through scenario analyses to prevent them before they occur. The application of simulation is suitable in this case, especially due to the high complexity that arises from the simultaneous consideration of material and energy flows. In contrast to other simulation studies in production, plants not only have to be modeled logistically, but data on their performance profiles for different states are also necessary.

The objective was to simulate a production schedule generated by the Manufacturing Execution System (MES) for this production area and to detect impending load peaks. In the event of a load peak, the simulation is able to determine alternatives and propose them to the production control system in order to set a load profile that does not cause additional costs in addition to the logistical target values. In addition to the consumption by the production, the grid consumption is also influenced by the self-generation by the photovoltaic system. If electricity is generated by the photovoltaic system, the production output can be increased without creating the risk of a load peak at the grid connection. Therefore, production must also be oriented to the generation forecast. The content of the simulation is shown in Fig. 7.8.

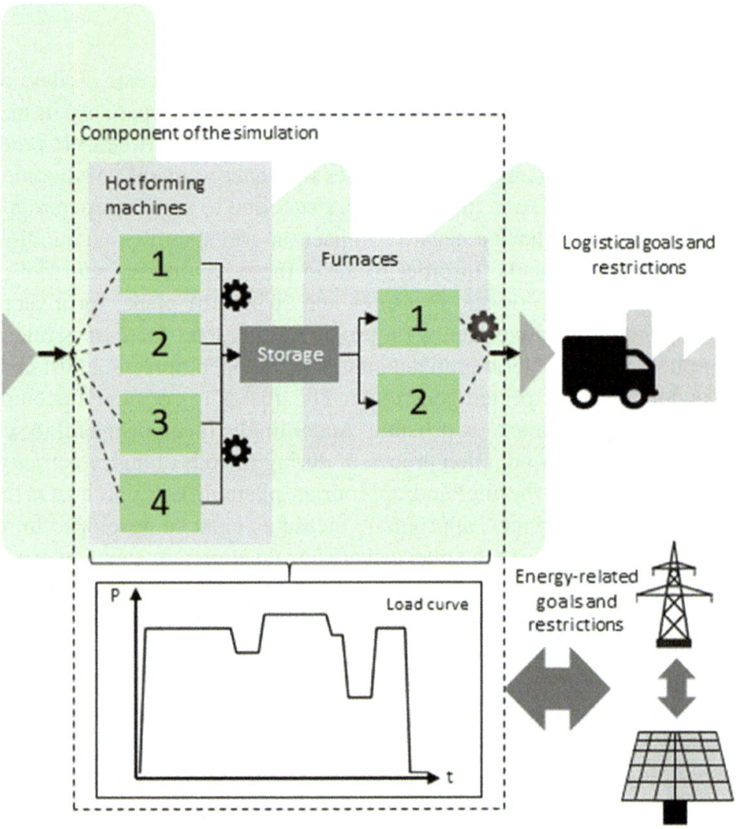

Fig. 7.8 System overview for the bevel gear fabrication

7.6.2.2 Procedure of the Simulation Project

The process follows the basic description of the VDI guideline for conducting a simulation study. Special attention should be paid in this case to the definition of the system boundary. It may be necessary to integrate facilities into the simulation that—while having no direct influence on the material flow—are worth a consideration due to their energy consumption. Furthermore, the recording of energy data should be mentioned. In this case, the data sources were the ERP system for the master data of the orders, as they are represented, e.g., in SAP or other tools, such as weight, geometry, and material. The production plan of the planning period was taken from the MES developed by a medium-sized software company, which also defined the horizon of the simulation. In addition, historical data from production were used to enable modeling of the plants. The energy data came from an energy management system. The simulation study was implemented with the help of the discrete event simulation software Tecnomatix Plant Simulation and a library developed by Siemens that enables a detailed representation of the energy consumption of plants. The library is currently still under development.

7.6.2.3 Modeling Application-Specific Aspects

The modeling has, of course, to represent the logical flow of material on the system under study. In addition, further elements have to be respected that model the energy consumption and links it with the material flow. To establish the model, as the first step, the available data are analyzed to identify these relationships. In order to align the continuous energy model with the discrete event logistics model, it is discretized. The model elements are further divided into parts with different energy characteristics, which are modeled separately.

Data analysis In order to gain an understanding of the relationships between production and energy consumption and to suitably represent the corresponding relationships in the simulation model, the production and energy data were analyzed. The aim was not only to be able to represent known product-machine combinations in the simulation application and then fall back on historical load curves, but to represent the actual causes. Regression analyses were used for the analysis, and it was possible to identify product characteristics that can essentially explain the subsequent power consumption of the machine.

Discretization The "E-Flex Tool", a library for Plant Simulation, is intended to enable the different energy states of the plants to be represented in greater detail than it is currently the case in Plant Simulation. The modeling of energy is currently represented by average power values for individual system states. For example, there is the operating state. Using average power values for long condition phases is not sufficient for forecasting a load profile, especially if load peaks are to be forecasted. The E-Flex tool further partitions the individual system states and also makes it possible to describe them as a function of different products. Thereby, the benefit of the E-Flex tool is that the energy states are subdivided into smaller steps during the discretization than before to gain a better approximation for the energy demand. This process is similar to integral calculus, where the area under a curve is approximated by subdividing simple rectangles further and further. In this way, it is possible to map the main factors influencing consumption. Furthermore, there is the possibility of scaling the operating states in order to be able to easily represent different operating modes of the systems (Prell et al. 2018, 2019) (Fig. 7.9).

In this application, product groups were formed that generate comparable load curves on the same machine due to their comparable masses and dimensions. The correlations between product characteristics and load were determined in previous data analyses. The representation of the load is still discrete, but this level of detail is sufficient depending on the application.

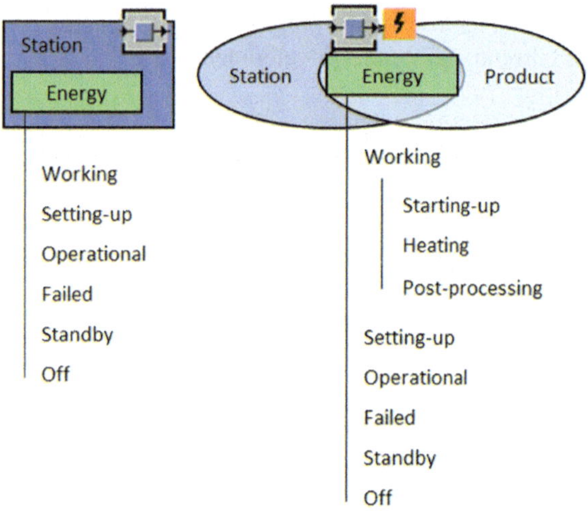

Fig. 7.9 Energy modeling

Modeling For modeling, corresponding aspects of energy must be considered, which may change the design of the models. The representation of the hot forming equipment will be presented as an example of this. The systems consist of two subsystems. In the first subsystem, the material is heated by an induction furnace. This is followed by forming through the press. The press has a volatile load curve due to the strokes, which leads to load peaks. The furnace, on the other hand, has a uniform load profile. To take this into account, the subsystems are each represented in the model by their own model modules in order to enable separate modeling of the energy requirement. Each subsystem in turn has a building block of the "E-Flex Tool" to describe the energetic profile (Fig. 7.10).

Simulation run For the simulation runs, production plans were used that were created by corresponding experts. The orders included in that production plans were divided into product groups according to the energy consumption characteristics found in the data analysis. The simulation was carried out with this modified production plan in which the specific orders were replaced with order groups. If the simulation found load peaks, different load avoidance scenarios were simulated and fed back to the

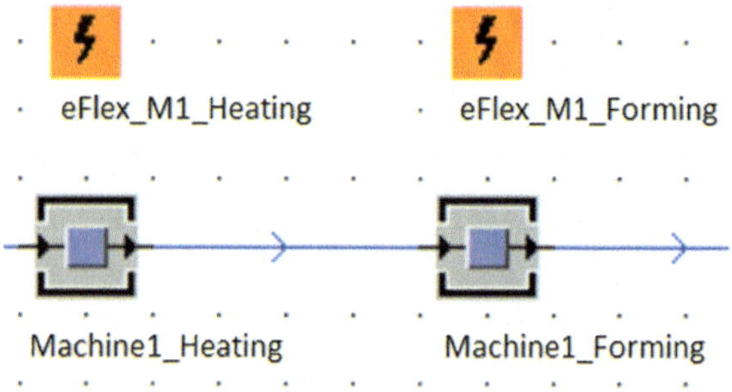

Fig. 7.10 Representation of a hot forming machine by two "SingleProc" blocks and their respective "E-Flex Tool" blocks (screenshot)

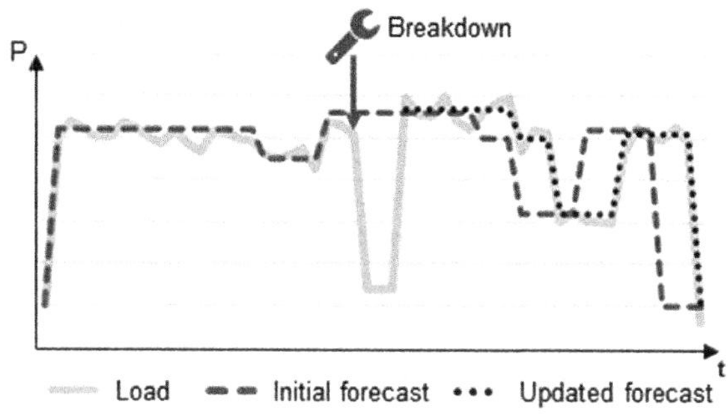

Fig. 7.11 Qualitative representation of the actual load curve and the simulation-based forecasts

production planning. The load can be adjusted by exchanging orders of different groups with each other, alternate the resource schedule or shifting production stops to different time windows.

7.6.2.4 Results of Experiments and Evaluation

The simulation studies made it possible to identify load peaks that would have inevitably occurred with the original production plan and to propose alternatives. In addition, the simulation was able to generate a load forecast whose forecast quality was good, despite the discretization. The forecast quality was measured using historical data. The deviation between the forecast and the observed power curve was analyzed graphically and by key figures like the root mean square error. The discrepancy between forecast and actual energy consumption was between 0.01% and 2.5% per forecast.

Since machine failures are not predicted in terms of time, they remain as undetected deviations in the load profile. To coordinate the restarting of failed machines and avoid creating a load peak through the start-up test, the simulation can be used to define a suitable start-up time window. In order to prevent an offset between the forecast and the actual load profile, the simulation should be run again after a longer production outage. In Fig. 7.11, the observed load profile and forecasts are shown.

The initial forecast fails as soon as the production breakdown occurs. For this reason, a new forecast is performed. It is clear that discretization is not able to perfectly match the load curve. Average load values per 15-min interval are decisive for the observation, and at this level of detail the forecast provides sufficient results.

7.6.2.5 Benefits and Lessons Learnt

Applying simulation achieves that the generation by the photovoltaic system, the power purchase via the grid, and the production are considered holistically. Taking into account the different influences, the risk of load peaks is analyzed. In case of imminent load peaks, measures are determined on the basis of the simulation that can prevent them.

Avoiding load peaks leads to the avoidance of non-inconsiderable costs in production and, thus, ensures the economic operation of production. As the simulation represents the production process and its energy flows in detail, further applications are conceivable. In the future, for example, the simulation model could be used to optimise the process according to the time-variable energy prices and, thus, further reduce energy costs. A further development into a digital twin could also make it possible to deal with unforeseen events more quickly and suggest appropriate interventions with a short reaction time.

7.7 Conclusions and Outlook

Due to the need to reduce greenhouse gas emissions and limit global warming, renewable energy sources are and will be—even more than now—an important pillar of the energy supply on a local and national level. In addition to the environmental benefits, actual cost savings may also be realized if renewables are used. However, the multitude of partially interrelated variables, such as energy demand, peak loads, weather effects, etc., entails a high complexity in the analysis and forecast of individual applications and their potential benefits. Therefore, simulation studies are carried out to address the complexity and analyze the impact of renewables on manufacturing sites, products, or even whole economies. The insights gained through these studies often include, how the total costs for electricity, the amount of GHG, or the utilization of manufacturing changes with varying energy supply by renewables. The need for such solutions will probably increase in the future due to the growing efforts for climate change mitigation. This trend gets an additional boost due to the current political situation, which pressures especially many European countries to get independent of imported fossil fuels as quickly as possible. This leads to more complex energy systems for manufacturing sites, which requires more-detailed simulation and control.

As has been shown, for example, in the two use cases, simulation models of interacting energy supply and manufacturing systems can be used not only to evaluate the economic (or environmental) viability of those systems, but also to develop and validate complex control strategies for manufacturing, renewables, and energy storages. These strategies can be used as tools to control actual manufacturing systems under constraints of renewable energy sources. A variety of similar examples can be found in the literature. Since the interest in energy-related studies in manufacturing has grown in the recent past, toolboxes have been developed to integrate energy modeling in simulation environments such as Plant Simulation to lower obstacles and make it easier to execute such studies. Additionally, an increasing share of IT tools for operations management, e.g., Manufacturing Execution Systems, now include tools to record and analyze energy consumption during manufacturing (Sauer et al. 2016). This will make the energy-related data of the demand side needed as input for simulation, such as state-based energy consumption for machines, more easily available, thus enhancing the practicality of these simulation studies.

References

Beier J (2017) Simulation approach towards energy flexible manufacturing systems. Springer International Publishing, Cham

Beier J, Neef B, Thiede S, Herrmann C (2016) Integrating on-site renewable electricity generation into a manufacturing system with intermittent battery storage from electric vehicles. Procedia CIRP 48:483–488

Binbin L, Yu T, Fred C, Tongdan J (2017) Toward net-zero carbon manufacturing operations: an onsite renewables solution. J Oper Res Soc 68(3):308–321

European Commission (2021) Energy and the green deal. https://ec.europa.eu/info/strategy/priorities-2019-2024/european-green-deal/energy-and-green-deal_en. Accessed 7 Aug 2021

European Financial Reporting Advisory Group (2022) European sustainability reporting standards. https://www.efrag.org/. Accessed 2 June 2022

Eurostat (2021) Share of renewable energy in gross final energy consumption. https://ec.europa.eu/eurostat/web/products-datasets/-/t2020_rd330. Accessed 7 Aug 2021

Gabel T (2009) Multi-agent reinforcement learning approaches for distributed job-shop scheduling problems. Ph.D. thesis, University Osnabrück, Mathematik/Informatik, Osnabrück

Gonzalez Ordiano JA, Waczowicz S, Reischl M, Mikut R, Hagenmeyer V (2017) Photovoltaic power forecasting using simple data-driven models without weather data. Comput Sci Res Develop 32(1–2):237–246

Hollmann M (2006) System dynamics modeling and simulation of distributed generation for the analysis of a future energy supply. In: Groessler A, Rouwette EAJA, Langer RS, Rowe JI, Yanni JM (eds) Proceedings of the international conference of the system dynamics society. Nijmegen, The Netherlands, pp 23–27

International Energy Agency (2020) Projected costs of generating electricity. 2020 Edition, International Energy Agency, Paris, France

International Energy Agency (2021) Share of renewables. https://www.iea.org/fuels-and-technologies/renewables. Accessed 7 Aug 2021

International Energy Agency, IRENA, UNSD, World Bank, WHO (2020) Tracking SDG7: the energy progress report 2020. World Bank, Washington DC, USA

IRENA (2021) Renewable power generation costs in 2020. International Renewable Energy Agency, Abu Dhabi, UAE

Jain S, Lindskog E, Johansson B (2012) Supply chain carbon footprint tradeoffs using simulation. In: Laroque C, Himmelspach J, Pasupathy R, Rose O, Uhrmacher AM (eds) Proceedings of the 2012 Winter Simulation Conference. IEEE, Piscataway, NJ, pp 3168–3179

Jeon C, Shih J (2014) Long-term renewable energy technology valuation using system dynamics and Monte Carlo simulation: photovoltaic technology case. Energy 66:447–457

Junge M (2007) Simulationsgestützte Entwicklung und Optimierung einer energieeffizienten Produktionssteuerung. Ph.D. thesis, University Kassel, Produktion & Energie, Kassel

Kaltschmitt M, Streicher W, Wiese A (2020) Erneuerbare Energien. Springer, Berlin Heidelberg

Kelly C, Onat N, Tatari O (2019) Water and carbon footprint reduction potential of renewable energy in the United States: a policy analysis using system dynamics. J Clean Prod 228:910–926

Kneiske TM, Hoffmann C (2014) Eigenstromnutzung. In: Neugebauer R (ed) Handbuch ressourcenorientierte Produktion. Carl Hanser, München Wien

Konstantin P (2013) Praxisbuch Energiewirtschaft: Energieumwandlung, -transport und -beschaffung im liberalisierten Markt. Springer Vieweg, Berlin Heidelberg van Laarhoven PJM, Aarts EH (1992) Simulated annealing. Theory and applications. Kluwer, Dordrecht

Materi S, D'Angola A, Enescu D, Renna P (2021) Reducing energy costs and CO_2 emissions by production system energy flexibility through the integration of renewable energy. Prod Eng Res Devel 15:667–681

Mazhari E, Zhao J, Celik N, Lee S, Son YJ, Head L (2011) Hybrid simulation and optimization-based design and operation of integrated photovoltaic generation, storage units, and grid. Simul Model Pract Theory 19(1):463–481

Moon YB (2016) Simulation modelling for sustainability: a review of the literature. Int J Sustain Eng 10(1):1–18

Oladeji AS, Akorede MF, Salihu A, Mohammed A (2021) Simulation-based optimization of hybrid renewable energy system for off-grid rural electrification. Int J Renew Energy Develop 10(4):667–686

Oneto L, Laureri F, Robba M, Delfino F, Anguita D (2018) Data-driven photovoltaic power production nowcasting and forecasting for polygeneration microgrids. IEEE Syst J 12(3):2842–2853

Prell B, Strehle HM, Schlegel A, Putz M (2019) Vergleichende Darstellung zweier Werkzeuge zur Simulation energetischer Aspekte in der Produktion. In: Putz M, Schlegel A (eds) Simulation in Produktion und Logistik 2019. Verlag Wissenschaftliche Skripten, Auerbach, pp 131–142

Prell B, Stoldt J, Putz M, Schlüter P, Szabo A, Bergs C, Unterberger E, Reinhart G, Moog D, Flum D, Abele E (2018) Simulation von Steuerungsmaßnahmen zum energieflexiblen Fabrikbetrieb. In: Proceedings Vernetzt Planen und Produzieren. Chemnitz, Germany, pp 79–88

RAY (2022) RLlib: Scalable, state of the art reinforcement learning. https://www.ray.io/rllib. Accessed 22 April 2022

Reddi KR, Weilin L, Wang B, Moon YB (2013) System dynamics modelling of hybrid renewable energy systems and combined heating and power generator. Int J Sustain Eng 6(1):31–47

Robalino-Lopez A, Mena-Nieto A, Garcia-Ramos JE (2014) System dynamics modeling for renewable energy and CO_2 emissions: a case study of Ecuador. Energy Sustain Dev 20(1):11–20

Roedger JM, Beier J, Schoenemann M, Schulze C, Thiede S, Bey N, Herrmann C, Hauschild MZ (2021) Combining life cycle assessment and manufacturing system simulation: evaluating dynamic impacts from renewable energy supply on product-specific environmental footprints. Int J Precision Eng Manuf-Green Technol 8:1007–1026

Roesch M, Linder C, Zimmermann R, Rudolf A, Hohmann A, Reinhart G (2020) Smart grid for industry using multi-agent reinforcement learning. Appl Sci 10(19):6900. https://doi.org/10.3390/app10196900

Saavedra MR, Fontes HOC, Freires FGM (2018) Sustainable and renewable energy supply chain: a system dynamics overview. Renew Sustain Energy Rev 82:247–259

Sanders L, Lopez S, Guzmann G, Jimenez J, Tongdan J (2012) Simulation of a green wafer fab featuring solar photovoltaic technology and storage system. In: Laroque C, Himmelspach J, Pasupathy R, Rose O, Uhrmacher AM (eds) Proceedings of the 2012 Winter Simulation Conference. IEEE, Piscataway, NJ, pp 1–12

Sauer A, Weckmann S, Zimmerman F (2016) Softwarelösungen für das Energiemanagement von morgen. https://www.eep.uni-stuttgart.de/dokumente/2016_EMS-Studie_Softwarlosungen-fuer-das-Energiemanagement-von-morgen.pdf. Accessed 11 Sept 2021

Schlegel A, Stoldt J, Putz M (2013) Erweiterte Integration energetischer Betrachtung in der Materialflusssimulation. In: Dangelmaier W, Laroque C, Klaas A (eds) Simulation in Produktion und Logistik 2013. HNI Verlagsschriftenreihe, Paderborn, Germany, pp 187–196

Schultz C, Sellmaier P, Reinhart G (2015) An approach for energy-oriented production control using energy flexibility. Procedia CIRP 29:197–202

Schulze C, Blume S, Siemon L, Herrmann C, Thiede S (2019) Towards energy flexible and energy self-sufficient manufacturing systems. Procedia CIRP 81:683–688

Shih YH, Tseng C-H (2014) Cost-benefit analysis of sustainable energy development using life-cycle co-benefits assessment and the system dynamics approach. Appl Energy 119:57–66

Taboada H, Xiong Z, Jin T, Jimenez J (2012) Exploring a solar photovoltaic-based energy solution for green manufacturing industry. IEEE international conference on automation science and engineering (CASE), Seoul, South Korea, pp 40–45

Tao C, Shanxu D, Changsong C (2010) Forecasting power output for grid-connected photovoltaic power system without using solar radiation measurement. In: The 2nd international symposium on power electronics for distributed generation systems. HeFei, China, pp 773–777

Thiede S (2012) Energy efficiency in manufacturing systems. Springer, Berlin Heidelberg

Thornton A, Kim SJ, Kara S (2018) Sizing a hybrid renewable energy system to reduce energy costs at various levels of robustness for an industrial site. Procedia CIRP 69:371–376

Tongdan J, Subramanyam VK, Castillo-Villar KK, Sun F (2020) Optimal sizing of renewable microgrid for flow shop systems under island operations. Procedia Manuf 51(1):1779–1784

Umweltbundesamt (2021) Erneuerbare Energie in Zahlen. https://www.umweltbundesamt.de/themen/klima-energie/erneuerbare-energien/erneuerbare-energien-in-zahlen#strom. Accessed 7 Aug 2021

United Nations (2019) The sustainable development goals report 2019. United Nations, New York, USA

United Nations (2021) Business and climate change: rising public awareness creates significant opportunity. https://www.un.org/en/chronicle/article/business-and-climate-change-rising-public-awareness-creates-significant-opportunity. Accessed 2 Dec 2022

Verein Deutscher Ingenieure (2014) VDI Guideline 3633 Simulation of systems in materials handling, logistics and production – Part 1. Fundamentals. Beuth, Berlin

Weinert N, Mose C (2016) Electrical load management for production equipment applying a decentralized optimization approach. Procedia CIRP 48:259–264

Weinert N, Chiotellis S, Seliger G (2011) Methodology for planning and operating energy-efficient production systems. CIRP Ann Manuf Technol 60:41–44

Woltmann S, Zarte M, Kittel J, Pechmann A (2018) Agent based simulation model of virtual power plants for greener manufacturing. Procedia CIRP 69:377–382

Xu R, Chen H, Sun X (2012) Short-term photovoltaic power forecasting with weighted support vector machine. In: IEEE international conference on automation and logistics. Zhengzhou, China, pp 248–253

Zahraee S, Assadi MK, Saidur R (2016) Application of artificial intelligence methods for hybrid energy system optimization. Renew Sustain Energy Rev 66:617–630

SPRINGER NATURE

GPSR Compliance

The European Union's (EU) General Product Safety Regulation (GPSR) is a set of rules that requires consumer products to be safe and our obligations to ensure this.

If you have any concerns about our products, you can contact us on ProductSafety@springernature.com

In case Publisher is established outside the EU, the EU authorized representative is:

Springer Nature Customer Service Center GmbH
Europaplatz 3
69115 Heidelberg, Germany

The manufacturer's authorised representative in the EU is Springer Nature Customer Service Centre GmbH, Europaplatz 3, 69115 Heidelberg, Germany. If you have any concerns regarding our products, please contact ProductSafety@springernature.com

Printed and bound by CPI Group (UK) Ltd, Croydon, CR0 4YY

01/10/2026

02225900-0001